VOLUME	EDITOR-IN-CHIEF	PAGES	
23	Lee Irvin Smith	124	*Out of print*
24	The Late Nathan L. Drake	119	*Out of print*
25	The Late Werner E. Bachmann	120	*Out of print*
26	The Late Homer Adkins	124	*Out of print*
27	R. L. Shriner	121	*Out of print*
28	H. R. Snyder	121	*Out of print*
29	Cliff S. Hamilton	119	*Out of print*

Collective Vol. 3 A revised edition of Annual
Volumes 20–29
E. C. Horning, *Editor-in-Chief*   890

30	The Late Arthur C. Cope	115	*Out of print*
31	R. S. Schreiber	122	
32	Richard T. Arnold	119	*Out of print*
33	Charles C. Price	115	*Out of print*
34	William S. Johnson	121	*Out of print*
35	T. L. Cairns	122	*Out of print*
36	N. L. Leonard	120	
37	James Cason	109	*Out of print*
38	John C. Sheehan	120	
39	Max Tishler	114	

Collective Vol. 4 A revised edition of Annual
Volumes 30–39
Norman Rabjohn, *Editor-in-Chief*   1036

40	Melvin S. Newman	114	
41	John D. Roberts	118	
42	Virgil Boekelheide	118	*Out of print*
43	B. C. McKusick	124	*Out of print*
44	William E. Parham	131	
45	William G. Dauben	118	
46	E. J. Corey	146	
47	William D. Emmons	140	
48	Peter Yates	164	
49	Kenneth B. Wiberg	124	

Collective Vol. 5 A revised edition of Annual
Volumes 40–49
Henry E. Baumgarten, *Editor-in-Chief*

50	Ronald Breslow	110
51	Richard E. Benson	147
52	Herbert O. House	139

NO LONGER PROPERTY OF
FALVEY MEMORIAL LIBRARY

ORGANIC SYNTHESES

CONTRIBUTORS

(Other than members of the Board)

Leona M. Baclawski	D. C. Muchmore
Richard F. Borch	M. Murakami
H. C. Brown	Martin Nilsson
L. Caglioti	E. J. O'Connell, Jr.
A. Calder	Roland Ohme
C. E. Castro	D. C. Owsley
J. C. Collins	P. J. Pearce
Ismael Colon	Charles J. Pedersen
D. S. Connor	Foster Pigott
Robert D. DeMaster	Helmut Preuschhof
A. R. Forrester	D. H. Richards
Martin Gall	M. R. Rifi
Gary W. Griffin	N. F. Scilly
George H. Hawks	Martin F. Semmelhack
Paul M. Helquist	Edward C. Taylor
S. P. Hepburn	G. N. Taylor
W. W. Hess	H.-J. Teuber
Hans-Ulrich Heyne	Ruth S. Wade
G. W. Klein	Pius A. Wehrli
William G. Kofron	Thomas N. Wheeler
T. A. Marolewski	N. C. Yang
Alexander McKillop	M. Yoshioka
Ferene Merenyi	G. Zweifel

ORGANIC SYNTHESES

AN ANNUAL PUBLICATION OF SATISFACTORY
METHODS FOR THE PREPARATION
OF ORGANIC CHEMICALS

VOLUME 52

1972

ADVISORY BOARD

C. F. H. ALLEN
RICHARD T. ARNOLD
HENRY E. BAUMGARTEN
A. H. BLATT
VIRGIL BOEKELHEIDE
T. L. CAIRNS
JAMES CASON
H. T. CLARKE
J. B. CONANT
E. J. COREY
WILLIAM G. DAUBEN
WILLIAM D. EMMONS
ALBERT ESCHENMOSER
L. F. FIESER
R. C. FUSON
HENRY GILMAN
C. S. HAMILTON
W. W. HARTMAN
E. C. HORNING

JOHN R. JOHNSON
WILLIAM S. JOHNSON
N. J. LEONARD
B. C. McKUSICK
C. S. MARVEL
MELVIN S. NEWMAN
C. R. NOLLER
W. E. PARHAM
CHARLES C. PRICE
NORMAN RABJOHN
JOHN D. ROBERTS
R. S. SCHREIBER
JOHN C. SHEEHAN
RALPH L. SHRINER
LEE IRVIN SMITH
H. R. SNYDER
MAX TISHLER
KENNETH B. WIBERG
PETER YATES

BOARD OF EDITORS

HERBERT O. HOUSE, *Editor-in-Chief*

RICHARD E. BENSON
RONALD BRESLOW
ARNOLD BROSSI
GEORGE H. BÜCHI

ROBERT E. IRELAND
SATORU MASAMUNE
JERROLD MEINWALD
WATARU NAGATA

WAYLAND E. NOLAND, *Secretary to the Board*
University of Minnesota, Minneapolis, Minnesota

FORMER MEMBERS OF THE BOARD, NOW DECEASED

ROGER ADAMS
HOMER ADKINS
WERNER E. BACHMANN
WALLACE H. CAROTHERS

ARTHUR C. COPE
NATHAN L. DRAKE
OLIVER KAMM
FRANK C. WHITMORE

JOHN WILEY AND SONS

NEW YORK · LONDON · SYDNEY · TORONTO

Copyright © 1972, by John Wiley & Sons, Inc.

All rights reserved. Published simultaneously in Canada.

No part of this book may be reproduced by any means, nor transmitted, nor translated into a machine language without the written permission of the publisher.

"John Wiley & Sons, Inc. is pleased to publish this volume of Organic Syntheses on behalf of Organic Syntheses, Inc. Although Organic Syntheses, Inc. has assured us that each preparation contained in this volume has been checked in an independent laboratory and that any hazards that were uncovered are clearly set forth in the write-up of each preparation, John Wiley & Sons, Inc. does not warrant the preparations against any safety hazards and assumes no liability with respect to the use of the preparations."

Library of Congress Catalog Card Number: 21-17747

ISBN 0-471-41560-X

Printed in the United States of America

10 9 8 7 6 5 4 3 2 1

NOMENCLATURE

Preparations appear in the alphabetical order of common names of the compounds or names of the synthetic procedures. For convenience in surveying the literature concerning any preparation through *Chemical Abstracts* subject indexes, the *Chemical Abstracts* indexing name for each compound is given as a subtitle if it differs from the common name used as the title.

SUBMISSION OF PREPARATIONS

Chemists are invited to submit for publication in *Organic Syntheses* procedures for the preparation of compounds that are of general interest, as well as procedures that illustrate synthetic methods of general utility. It is fundamental to the usefulness of *Organic Syntheses* that submitted procedures represent optimum conditions, and the procedures should have been checked carefully by the submitters, not only for yield and physical properties of the products, but also for any hazards that may be involved. Full details of all manipulations should be described, and the range of yield should be reported rather than the maximum yield obtainable by an operator who has had considerable experience with the preparation. For each solid product the melting-point range should be reported, and for each liquid product the range of boiling point and refractive index should be included. In most instances, it is desirable to include additional physical properties of the product, such as ultraviolet, infrared, mass, or nuclear magnetic resonance spectra, and criteria of purity such as gas chromatographic data. In the event that any of the reactants are not readily commercially available at reasonable cost, their preparation should be described in as complete detail and in the same manner as the preparation of the product of major interest. The sources of the reactants should be described in notes, and the physical properties (such as boiling point, index of refraction, melting point) of the reactants should be included except where standard commercial grades are specified.

Beginning with Volume 49, Sec. 3., Method of Preparation, and Sec. 4., Merits of the Preparation, have been combined into a single new Sec. 3., Discussion. In this section should be described other practical methods for accomplishing the purpose of the procedure that have appeared in the literature. It is unnecessary to mention methods that have been published but are of no practical synthetic value. Those features of the procedure that recommend it for publication in *Organic Syntheses* should be cited (synthetic method of considerable scope, specific compound of interest not likely to be made available commercially, method that gives better yield or is less laborious than other methods, etc.). If possible, a brief discussion of the scope and limitations of the procedure as applied to other examples as well as a comparison of the method with the other methods cited should be included. If necessary to the understanding or use of the method for related syntheses, a brief discussion of the mechanism may be placed in this section. The present emphasis of *Organic Syntheses* is on model procedures rather than on specific compounds (although the latter are still welcomed), and the Discussion section should be written to help the reader decide whether and how to use the procedure in his own research. Three copies of each procedure should be submitted to the Secretary of the Editorial Board. It is sometimes helpful to the Board if there is an accompanying letter setting forth the features of the preparations that are of interest.

Additions, corrections, and improvements to the preparations previously published are welcomed and should be directed to the Secretary.

PREFACE

This volume of *Organic Syntheses* continues the policy of recent volumes in this series to emphasize detailed descriptions that can serve as model experimental procedures for useful synthetic methods. A number of procedures for the synthesis of specific reagents of special interest are also included. A sizable fraction of the model experimental procedures were chosen to illustrate an increasingly important group of synthetic methods that utilize organometallic compounds as reactants or as intermediates.

Among the methods involving organometallic reactants or intermediates, three procedures make use of organoaluminum intermediates. DIETHYLALUMINUM CYANIDE, generated either separately or in the reaction mixture, is used to form a cyanohydrin that is dehydrated to 1-CYANO-6-METHOXY-3,4-DIHYDRONAPHTHALENE and is used in a stereospecific enone hydrocyanation reaction to form 3β-ACETOXY-5α-CYANO-CHOLESTAN-7-ONE. Uses of organocopper(I) intermediates are illustrated in a coupling reaction with copper(I) phenylacetylide to form 2-PHENYLFURO[3,2-*b*]PYRIDINE and by the stereospecific conjugate addition of lithium dimethylcuprate to an enone followed by phosphorylation of the intermediate enolate and subsequent reductive cleavage to 5-METHYLCOPROST-3-ENE. The generation of specific lithium enolates, followed by alkylation, is also illustrated in the preparations of 2-BENZYL-2-METHYLCYCLOHEXANONE and 2-BENZYL-6-METHYLCYCLOHEXANONE. The preparation of METHALLYLBENZENE provides a model for the coupling of π-allylnickel halides with aryl halides and the preparation of 3-BUTYL-2-METHYLHEPT-1-EN-3-OL serves as an example of the generation of an organolithium compound in the presence of a reactant containing a carbonyl group. The selective metalation of a 2-methylpyridine derivative is exemplified by the formation of ETHYL 6-METHYLPYRIDINE-2-ACETATE, and the hydrogenolysis of an aryl bromide with the ethylenediamine complex of chromium(II) is illustrated by the conversion of 1-bromonaphthalene to NAPHTHALENE. Finally, the preparation of ISOPINOCAMPHEOL provides a model for the hydroboration of an olefin followed by oxidation to form an alcohol.

Other general synthetic methods illustrated include the oxidation of a primary alcohol to 1-HEPTANAL with the chromium trioxide-pyridine complex, the formation and photochemical Wolff rearrangement of an α-diazo ketone to form the D-NORANDROST-5-EN-3β-OL-16-CARBOXYLIC ACIDS, the use of sodium cyanoborohydride to reduce an iminium salt to N,N-DIMETHYLCYCLOHEXYLAMINE, and the use of a tosylhydrazone intermediate to reduce a ketone function to a methylene group in ANDROSTAN-17β-OL. Two electrochemical procedures provide examples of a controlled-potential reduction to form 1,1-BIS(BROMOMETHYL)CYCLOPROPANE and an electrolytic oxidation to form the NITROSODISULFONATE RADICAL. This radical (or Fremy's salt) is used in the selective oxidation of phenols to TRIMETHYL-p-BENZOQUINONE or to 4,5-DIMETHYL-1,2-BENZOQUINONE. The formation of ozonides and their reduction with sulfur dioxide to form acetals is exemplified by the preparation of 4H-1,4-THIAZINE 1,1-DIOXIDE and a general route to azoalkanes is illustrated by the preparation of AZOETHANE. Two preparative routes to cyclopropanes include the photolysis of iodoform in the presence of an olefin to form 2,2,3,3-TETRAMETHYLIODOCYCLOPROPANE and the successive treatment of a 1,5-diketone with iodine and with base to form *trans*-1,2-DIBENZOYLCYCLOPROPANE. The use of a cyclic acid anhydride in the Friedel-Crafts acylation of an olefin is used to prepare 2-ACETYLCYCLOPENTANE-1,3-DIONE.

The synthesis of specific compounds that are of general interest includes DIPHENYL KETENE; BENZYL CHLOROMETHYL ETHER, a useful alkylating agent; and 2-METHYL-2-NITROSOPROPANE, an excellent scavenger for free-radical intermediates. Also described are the syntheses of DIBENZO-18-CROWN-6 POLYETHER and DICYCLOHEXYL-18-CROWN-6 POLYETHER, two macrocyclic polyethers that form complexes with the cations of alkali metal salts.

The Board of Editors is grateful to the contributors of the preparations included in this volume and welcomes both the submission of preparations for future volumes and suggestions for change that will improve the usefulness of *Organic Syntheses*. The attention of submitters of preparations is drawn to the instructions on pages v and vi that describe the types of preparations we wish to obtain and also the information to be incorporated in each preparation.

In the continuation of a practice initiated with Volume 50 of *Organic Syntheses*, at the end of this volume appears an insert listing the preparations that have been received during the preceding year. These unchecked procedures, available from the Secretary's office for a nominal fee, allow the experimental procedures we receive to be accessible more rapidly to interested users. All preparations accepted by the Board of Editors will, of course, continue to be checked before final publication.

The Editor-in-Chief thanks the many contributors who responded to his request for preparations illustrating the use of organometallic reagents, and Mrs. Ann F. Edwards for her many contributions in the preparation and editing of the manuscript for this volume.

HERBERT O. HOUSE

Atlanta, Georgia
May 1972

CONTENTS

2-Acetylcyclopentane-1,3-dione	1
Aldehydes from Primary Alcohols by Oxidation with Chromium Trioxide: 1-Heptanal	5
Azoethane	11
Benzyl Chloromethyl Ether	16
3-Butyl-2-methylhept-1-en-3-ol	19
Controlled-Potential Electrolytic Reduction: 1,1-Bis(bromomethyl)cyclopropane	22
1,2-Diaroylcyclopropanes: trans-1,2-Dibenzoylcyclopropane	33
Diphenylketene	36
The Formation and Alkylation of Specific Enolate Anions from an Unsymmetrical Ketone: 2-Benzyl-2-methylcyclohexanone and 2-Benzyl-6-methylcyclohexanone	39
Formation and Photochemical Wolff Rearrangement of Cyclic α-Diazo Ketones: D-Norandrost-5-en-3β-ol-16-carboxylic Acids	53
Hydroboration of Olefins: (+)-Isopinocampheol	59
Hydrogenolysis of Carbon-halogen Bonds with Chromium(II)-en Perchlorate: Naphthalene from 1-Bromonaphthalene	62
Macrocyclic Polyethers: Dibenzo-18-crown-6 Polyether and Dicyclohexyl-18-crown-6 Polyether	66
Metalation of 2-Methylpyridine Derivatives: Ethyl 6-Methylpyridine-2-acetate	75
2-Methyl-2-nitrosopropane and its Dimer	77
Oxidation with the Nitrosodisulfonate Radical. I. Preparation and Use of Disodium Nitrosodisulfonate: Trimethyl-p-benzoquinone	83
II. Use of Dipotassium Nitrosodisulfonate (Fremy's Salt): 4,5-Dimethyl-1,2-benzoquinone	88
Preparation of Cyano Compounds Using Alkylaluminum Intermediates. I. Diethylaluminum Cyanide	90
II. 1-Cyano-6-methoxy-3,4-dihydronaphthalene	96
III. 3β-Acetoxy-5α-cyanocholestan-7-one	100
Preparation and Reductive Cleavage of Enol Phosphates: 5-Methylcoprost-3-ene	109
Reaction of Aryl Halides with π-Allylnickel Halides: Methallylbenzene	115

Reduction of Ketones By Use of the Tosylhydrazone
  Derivatives: Androstan-17-β-ol 122
Reductive Amination with Sodium Cyanoborohydride:
  N,N-Dimethylcyclohexylamine 124
Substitution of Aryl Halides with Copper(I) Acetylides:
  2-Phenylfuro[3,2-b]pyridine 128
2,2,3,3-Tetramethyliodocyclopropane 132
4H-1,4-Thiazine 1,1-Dioxide. 135

Subject Index 141

ORGANIC SYNTHESES

2-ACETYLCYCLOPENTANE-1,3-DIONE

(2–Acetyl-3-hydroxycyclopent-2-en-1-one, Note 1)

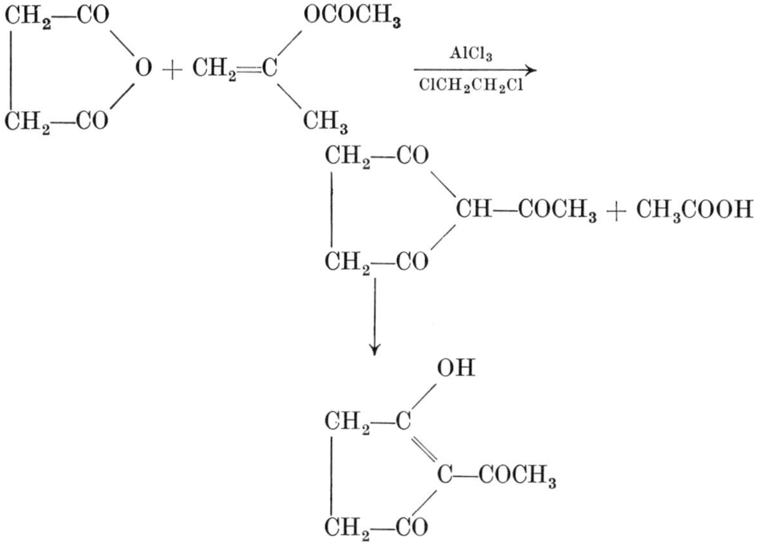

$$\underset{\text{CH}_2-\text{CO}}{\overset{\text{CH}_2-\text{CO}}{\diagdown}}\text{O} + \text{CH}_2=\underset{\text{CH}_3}{\overset{\text{OCOCH}_3}{\diagup}}\text{C} \xrightarrow[\text{ClCH}_2\text{CH}_2\text{Cl}]{\text{AlCl}_3}$$

$$\underset{\text{CH}_2-\text{CO}}{\overset{\text{CH}_2-\text{CO}}{\diagdown}}\text{CH}-\text{COCH}_3 + \text{CH}_3\text{COOH}$$

$$\downarrow$$

$$\underset{\text{CH}_2-\text{CO}}{\overset{\text{CH}_2-\text{C}}{\diagdown}}\overset{\text{OH}}{\underset{\diagdown}{\diagup}}\text{C}-\text{COCH}_3$$

Submitted by Ferenc Merényi and Martin Nilsson[1]
Checked by E. J. Corey, Joel I. Shulman, and Lawrence Libit

1. Procedure

Caution! Since hydrogen chloride is evolved in this reaction, it should be conducted in a hood.

In a dry 3-l. three-necked round-bottomed flask, equipped with a sealed mechanical stirrer (all-glass or glass-Teflon), a reflux condenser fitted with a calcium chloride drying tube and a 100-ml. dropping funnel, are placed 50.0 g. (0.500 mole) of finely powdered succinic anhydride (Note 2), 133.4 g. (1.00 mole) of freshly crushed anhydrous aluminum chloride (Note 3), and 500 ml. of anhydrous 1,2-dichloroethane. The mixture is stirred vigorously at room temperature for about 2 hours to dissolve as much as possible of the solid reactants. Then 50.0 g. (0.500 mole) of isopropenyl acetate (Note 4) is added rapidly through the dropping funnel; the reaction starts immediately as indicated by a rise in temperature to about 60° to 70°. The mixture is then refluxed for 15 minutes with continuous stirring. The hot reaction mixture, which contains a sticky oil, is poured into a stirred mixture of 200 ml. of aqueous 12M hydrochloric acid and 1000 g. of crushed ice, and the reaction flask is rinsed with part of the acidic aqueous phase. When the dark mass has dissolved, 200 ml. of concentrated aqueous hydrochloric acid is added, and the mixture is stirred vigorously for about 3 hours (Note 5). The dichloroethane phase is separated, and the aqueous phase is extracted with eight 600-ml. portions of dichloromethane. This extract is combined with the dichloroethane phase (Note 6) and extracted first with 600 ml. and then twice with 200-ml. portions of aqueous saturated sodium bicarbonate solution. The combined sodium bicarbonate extracts are washed with 200 ml. of dichloromethane and then cautiously acidified in a 3-l. beaker with 150 ml. of concentrated hydrochloric acid with vigorous stirring. The acidic solution (Note 7) is extracted first with 600 ml. and then with four 400-ml. portions of dichloromethane. The bulk of the dichloromethane is removed by distillation at atmospheric pressure with a water bath. The last 150 ml. of solvent is removed below room temperature under reduced pressure (about 20 mm.) to leave 24–27 g. of crude 2-acetylcyclopentane-1,3-dione as a light brown solid, m.p. 68–71° (Note 8).

The crude product is purified by decolorization with charcoal and recrystallization from 100–150 ml. of diisopropyl ether to give about 19–21 g. (27–30%) of colorless needles, m.p. 70–72°. This material is sufficiently pure for most purposes. Further purification may be achieved by further recrystallization from diisopropyl ether and/or sublimation at 60° (0.1 mm.) onto a cold-finger condenser to give material melting at 73–74° (Note 9).

2. Notes

1. 2-Acetylcyclopentane-1,3-dione is completely enolized in the solid state as well as in solution.[2] Indirect evidence indicates that the carbon-carbon double bond of the enol is within the ring.

2. The submitters used a practical grade of succinic anhydride obtained from Matheson Coleman and Bell.

3. The submitters used anhydrous, sublimed aluminum chloride obtained from E. Merck AG, Darmstadt, Germany. The checkers used analytical reagent grade material obtained from Mallinckrodt Chemical Works.

4. The submitters used either a pure grade (more than 98.5%) of isopropenyl acetate or a practical grade (95–98%) from Fluka AG, Buchs SG, Switzerland. Generally, it was not essential to distill the material before use.

5. This treatment increases the amount of product which can be extracted from the aqueous solution.

6. Evaporation of the solvents at this stage gives an oily product which may be decolorized with charcoal and then recrystallized from diisopropyl ether to give 2-acetylcyclopentane-1,3-dione, m.p. 68–71°. However, the extraction with sodium bicarbonate is preferred.

7. Decolorization of the acidic aqueous solution with charcoal at this stage improves the quality of the product.

8. 2-Acetylcyclopentane-1,3-dione is quite volatile and appreciable losses may occur if the evaporation of the concentrated solution is continued at elevated temperature.

9. Further amounts of 2-acetylcyclopentane-1,3-dione can be obtained by continuous extraction of the original aqueous phase with dichloromethane. The submitters have obtained total yields of product as high as 41–43%. 2-Acetylcyclopentane-1,3-dione is sparingly soluble in ether and is not extracted very efficiently from aqueous solution with ether.

3. Discussion

2-Acetylcyclopentane-1,3-dione has been obtained in small amounts from the aluminum chloride-catalyzed reactions of vinyl acetate and succinyl chloride in 1,1,2,2-tetrachloroethane.[5] 2-Acetylcyclopentane-1,3-diones have been prepared via Dieckmann condensation of 1,4-bisethoxycarbonylhexane-3,5-diones.[6] The present procedure is essentially that of Merényi and Nilsson[3] with some modifications.

The diacylation of isopropenyl acetate with anhydrides of dicarboxylic acids is applicable for the synthesis of several other cyclic β-triketones in moderate yield.[3,4] It has been used for the synthesis of 2-acetylcyclohexane-1,3-dione (40% yield), 2-acetyl-4-methylcyclopentane-1,3-dione (10% yield), 2-acetyl-4,4-dimethylcyclopentane-1,3-dione (10% yield), 2-acetyl-5,5-dimethylcyclohexane-1,3-dione (10% yield), 2-acetylcycloheptane-1,3-dione (12% yield) and 2-acetylindane-1,3-dione (25% yield). Maleic anhydrides under more drastic conditions give acetylcyclopent-4-ene-1,3-diones in yields from 5% to 12%.[7] The corresponding acylation of the enol acetate of 2-butanone with succinic anhydride has been used to prepare 2-methylcyclopentane-1,3-dione, an important intermediate in steroid synthesis.[8,9]

2-Acetylcyclopentane-1,3-dione can be hydrolyzed by aqueous $2M$ hydrochloric acid (2 days, 100°) to give cyclopentane-1,3-dione in 63% yield.[4]

1. Department of Organic Chemistry, Chalmers University of Technology and the University of Göteborg, S-4020 Göteborg 5, Sweden.
2. S. Forsén, F. Merényi, and M. Nilsson, *Acta Chem. Scand.*, **18**, 1208 (1964).
3. F. Merényi and M. Nilsson, *Acta Chem. Scand.*, **17**, 1801 (1963).
4. F. Merényi and M. Nilsson, *Acta Chem. Scand.*, **18**, 1368 (1964).
5. A. Sieglitz and O. Horn, *Chem. Ber.*, **84**, 607 (1951).
6. M. Vandewalle, *Bull. Soc. Chim. Belg.*, **73**, 628 (1964).
7. M. Nilsson, *Acta Chem. Scand.*, **18**, 441 (1964).
8. H. Scheck, G. Lehman, and G. Hilgetag, *J. Prakt. Chem.*, **35**, 28 (1967); *Angew. Chem.*, **79**, 97 (1967).
9. V. J. Grenda, G. W. Lindberg, N. L. Wendler, and S. H. Pines, *J. Org. Chem.*, **32**, 1236 (1967).

ALDEHYDES FROM PRIMARY ALCOHOLS BY OXIDATION WITH CHROMIUM TRIOXIDE: 1-HEPTANAL

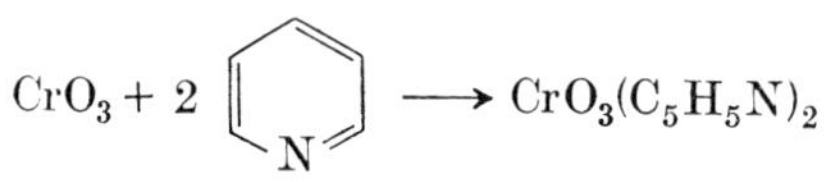

$$CrO_3 + 2\ C_5H_5N \longrightarrow CrO_3(C_5H_5N)_2$$

$$CH_3(CH_2)_5CH_2OH \xrightarrow[CH_2Cl_2,\ 25°]{CrO_3(C_5H_5N)_2} CH_3(CH_2)_5CHO$$

Submitted by J. C. COLLINS[1] and W. W. HESS[2]
Checked by R. T. UYEDA and R. E. BENSON

1. Procedure

A. *Preparation of Dipyridine Chromium(VI) Oxide* (Note 1). Caution! The reaction of chromium trioxide with pyridine is extremely exothermic; the preparation should be conducted in a hood, observing the precautions noted.

A dry, 1-l. three-necked flask, fitted with a sealed mechanical stirrer, a thermometer, and a drying tube, is charged with 500 ml. of anhydrous pyridine (Note 2). The pyridine is stirred and cooled to approximately 15° (Note 3) with an ice bath. The drying tube is periodically removed and 68 g. (0.68 mole) of anhydrous chromium(VI) oxide (Note 4) is added in portions through the neck of the flask over a 30-minute period. The chromium trioxide should be added at such a rate that the temperature does not exceed 20° and in such a manner that the oxide mixes rapidly with the pyridine and does not adhere to the side of the flask below the neck (Note 5). As the chromic oxide is added, an intensely yellow, flocculent precipitate separates from the pyridine and the viscosity of the mixture increases. When the addition is complete, the mixture is allowed to warm slowly to room temperature with stirring. Within one hour the viscosity of the mixture decreases and the initially yellow product changes to a deep red macrocrystalline form that settles to the bottom of the flask when stirring is discontinued. The supernatant pyridine is decanted from the complex and the crystals are washed several times by decantation with 250-ml. portions of anhydrous petroleum ether. The product is collected by filtration on a sintered glass funnel and washed with anhydrous petroleum ether, avoiding contact with the atmosphere as much as possible. The complex is dried at 10 mm. until it is free-flowing to leave 150–160 g. (85–91%) of dipyridine chromium(VI) oxide[3] as red crystals. The product is extremely hygroscopic; contact with moisture converts it rapidly to the yellow dipyridinium dichromate.[4] It is stored at 0° in a brown bottle (Note 6).

B. *General Oxidation Procedure for Alcohols.* A sufficient quantity of a 5% solution of dipyridine chromium (VI) oxide (Note 1) in anhydrous dichloromethane (Note 7) is prepared to provide a sixfold molar ratio of complex to alcohol. This excess is usually required for complete oxidation to the aldehyde. The freshly prepared, pure complex dissolves completely in dichloromethane at 25° at 5% concentration to give a deep red solution, but solutions usually contain small amounts of brown, insoluble material when prepared from crude complex (Note 8). The alcohol, either pure or as a solution in anhydrous methylene chloride, is added to the red solution in one portion with stirring at room temperature or lower. The oxidation of unhindered primary (and secondary) alcohols proceeds to completion within 5 minutes to 15 minutes at 25° with deposition of brownish-black, polymeric, reduced chromium-pyridine products (Note 9). When deposition of reduced chromium compounds is complete (monitoring the reaction by gas chromatography or thin-layer chromatography analysis is helpful), the supernatant liquid is decanted from the (usually tarry) precipitate and the precipitate is rinsed thoroughly with dichloromethane (Note 10).

The combined dichloromethane solutions may be washed with dilute hydrochloric acid, sodium bicarbonate solution, and water to remove excess traces of pyridine and chromium salts, or they may be filtered directly through a filter aid or passed through a chromatographic column. The product is obtained by removal of dichloromethane; any pyridine that remains can often be removed under reduced pressure.

C. *Illustrative Example: 1-Heptanal.* A dry, 1-l. three-necked round-bottomed flask is equipped with a mechanical stirrer, and 650 ml. of anhydrous dichloromethane (Note 7) is added. Stirring is begun and 77.5 g. (0.30 mole) of dipyridine chromium(VI) oxide (Note 1) is added at room temperature. To the resulting solution is added 5.8 g. (0.050 mole) of 1-heptanol (Note 11) in one portion. After stirring for 20 minutes, the supernatant solution is decanted from the insoluble brown gum. The insoluble residue is washed with three 100-ml. portions of ether and the ether and methylene chloride solutions are combined. The resulting solution is washed successively with 300 ml. of aqueous 5% sodium hydroxide solution, with 100 ml. of aqueous 5% hydrochloric acid solution (Note 12), with two 100-ml. portions of saturated aqueous sodium bicarbonate solution, and, finally, with 100 ml. of saturated aqueous sodium chloride solution. The organic layer is dried over anhydrous magnesium sulfate, and the solvent is removed by distillation. Distillation of the residual oil at reduced pressure through a small Claisen head separates 4.0–4.8 g. (70–84%) of 1-heptanal, b.p. 80–84° (65 mm.), n^{25} D 1.4094 (Note 13).

2. Notes

1. Dipyridine chromium(VI) oxide is available from Eastman Organic Chemicals. To be an effective reagent, it must be anhydrous. It should form a red solution on dissolution in anhydrous methylene chloride.

2. Commercial reagent-grade pyridine was used. The checkers used material available from Allied Chemical Corporation, B and A grade.

3. To avoid the accumulation of excess unchanged chromium trioxide, and rapid temperature rise when it does react, *the initial temperature of the pyridine should never be below* 10°.

4. Reagent-grade chromium(VI) oxide was dried over phosphorus pentoxide. The checkers used material available from Allied Chemical Corporation, B and A grade.

5. A glassine paper cone or glass funnel inserted in the drying tube neck of the flask during additions proved satisfactory, provided the cone or funnel was replaced frequently.

The paper must be discarded carefully, since it may inflame. Adding the chromium trioxide from a flask through rubber tubing proved dangerous because it caused local excesses of the oxide below and in the neck of the flask. Pyridine added to chromium trioxide spontaneously ignites causing spot fires that extinguish themselves rapidly if the pyridine temperature is below 20° and stirring is efficient. Such fires should and can be avoided.

6. Since the complex itself loses pyridine under reduced pressure and darkens with surface decomposition, it should not be stored under vacuum or over acidic drying agents. Minimal exposure to the atmosphere is required to prevent hydration of the complex. The checkers found that a free-flowing product was obtained on drying for one hour.

7. Commercial dichloromethane was dried by distillation from phosphorus pentoxide. The dichloromethane may also be decanted from phosphorus pentoxide prior to use. Small amounts of suspended phosphorus pentoxide do not seem to interfere with the oxidation.

8. If the complex does not dissolve in dichloromethane to form a red solution, either the complex has been hydrated in handling, or the dichloromethane is not anhydrous.

9. After the alcohol and complex are thoroughly mixed, the mixture may be stirred near its surface to avoid fouling of the stirrer by the thick, chromium-containing reduction product. Alternatively, the mixture may be swirled periodically to collect the reduction product on the side of the flask.

10. The reduced chromium precipitate is soluble in saturated sodium bicarbonate solution but no additional aldehyde was obtained on extracting this bicarbonate solution with ether.

11. 1-Heptanol, obtained from Aldrich Chemical Company, Inc., was distilled before use, b.p. 176°.

12. A second washing with 100 ml. of aqueous 5% hydrochloric acid solution will reduce the amount of pyridine present in the final product without significantly decreasing the yield.

13. The product shows a strong band at 1720 cm.$^{-1}$ (C=O) in the infrared. Gas chromatographic analysis indicated a purity of about 94–98%, with pyridine as the major impurity.

3. Discussion

Chromic acid, in a variety of acidic media, has been used extensively for the oxidation of primary alcohols to aldehydes but rarely has provided aldehydes in greater than 50% yield.[5] Chromium trioxide in pyridine was introduced as a unique, nonacidic reagent for alcohol oxidations and has been used extensively to prepare ketones,[6] but has been applied with only limited success to the preparation of aldehydes. While o-methoxybenzaldehyde was obtained in 89% yield, 4-nitrobenzaldehyde and n-heptanal were obtained in 28% and 10% yields, respectively.[7]

Using the preformed dipyridine chromium(VI) oxide in dichloromethane, the rate of chromate ester formation and decay to the aldehyde[8] is enhanced at least twentyfold over the rate observed in pyridine solution.[4] Isolation of products is facile and aldehydes appear to be relatively stable to excess reagent. Relatively few applications of this reagent have been reported, but the method has provided virtually quantitative yields of aldehyde-intermediates in the synthesis of prostaglandins[9] and steroids.[10] Another paper reports the oxidation of 2-vinylcyclopropylcarbinol to the aldehyde in 85% yield.[11] Although excess reagent is required for the oxidations (usually sixfold), the reaction conditions are so mild and isolation of products so easy that the complex will undoubtedly find broad use as a specialty reagent. Isolation of the complex can be avoided by *in situ* preparation of the chromium oxide/pyridine complex.[12]

Other general syntheses of aldehydes from primary alcohols involve the use of dimethyl sulfoxide[13] with a dehydrating agent such as dicyclohexylcarbodiimide and phosphoric acid (or pyridinium trifluoroacetate),[14] diethylcarbodiimide,[15] or sulfur trioxide.[16] Alternatively, dimethyl sulfoxide has been used with derivatives of the alcohol such as the chloroformate,[17] the iodide,[18] and the tosylate.[19] Tertiary butyl chromate[20] and lead tetraacetate in pyridine[21] have been employed to oxidize aliphatic primary alcohols to aldehydes while manganese dioxide[22] has been used to prepare aromatic and α,β-unsaturated aldehydes.

1. Chemistry Division, Sterling-Winthrop Research Institute, Rensselare, New York 12144.
2. Department of Chemistry, Illinois Wesleyan University, Bloomington, Illinois 61701.
3. H. H. Sisler, J. D. Bush, and O. E. Accountius, *J. Amer. Chem. Soc.*, **70**, 3827 (1948).
4. J. C. Collins, W. W. Hess, and F. J. Frank, *Tetrahedron Lett.*, 3363 (1968).
5. K. B. Wiberg, "Oxidation in Organic Chemistry," Part A, Academic Press, New York, 1965, p. 142; J. Carnduff, *Quart. Rev. (London)*, **20**, 169 (1966).
6. G. I. Poos, G. E. Arth, R. E. Beyler, and L. H. Sarett, *J. Amer. Chem. Soc.*, **75**, 422 (1953).
7. J. R. Holum, *J. Org. Chem.*, **26**, 4814 (1961).
8. W. Nagata, T. Wakabayashi, Y. Hayase, M. Narisada, and S. Kamata, *J. Amer. Chem. Soc.*, **92**, 3202 (1970); **93**, 5740 (1971).
9. E. J. Corey, N. M. Weinshenker, T. K. Schaaf, and W. Huber, *J. Amer. Chem. Soc.*, **91**, 5675 (1969).
10. W. S. Johnson, C. A. Harbert, and R. D. Stipanovic, *J. Amer. Chem. Soc.*, **90**, 5279 (1968).
11. A. W. Burgstahler and C. M. Groginsky, *Trans. Kans. Acad. Sci.*, **72**, 486 (1969).
12. R. Ratcliffe and R. Rodehorst, *J. Org. Chem.*, **35**, 4000 (1970).
13. W. W. Epstein and F. W. Sweat, *Chem. Rev.*, **67**, 247 (1967).
14. K. E. Pfitzner and J. G. Moffatt, *J. Amer. Chem. Soc.*, **85**, 3027 (1963); *J. Amer. Chem. Soc.*, **87**, 5670 (1965).
15. A. F. Cook and J. G. Moffatt, *J. Amer. Chem. Soc.*, **89**, 2697 (1967).
16. J. R. Parikh and W. von E. Doering, *J. Amer. Chem. Soc.*, **89**, 5505 (1967).
17. D. H. R. Barton, B. J. Garner, and R. H. Wightman, *J. Chem. Soc.*, 1855 (1964).
18. A. P. Johnson and A. Pelter, *J. Chem. Soc.*, 520 (1964).
19. N. Kornblum, W. J. Jones, and G. J. Anderson, *J. Amer. Chem. Soc.*, **81**, 4113 (1959).
20. R. V. Oppenauer and H. Oberrauch, *An. Asoc. Quim. Argent.*, **37**, 246 (1949) [*C. A.*, **44**, 3871 (1950)]; H.-W. Bersch and A. v. Mletzko, *Arch. Pharm. (Weinheim)*, **291**, 91 (1958); T. Suga, K. Kihara, and T. Matsuura, *Bull. Chem. Soc. Jap.*, **38**, 893 (1965); T. Suga and T. Matsuura, *Bull. Chem. Soc. Jap.*, **38**, 1503 (1965).
21. R. E. Partch, *Tetrahedron Lett.*, 3071 (1964).
22. R. M. Evans, *Quart. Rev. (London)*, **13**, 61 (1959).

AZOETHANE

$$C_2H_5NH_2 + SO_2Cl_2 \xrightarrow[-15°]{\text{pyridine, petroleum ether}} (C_2H_5NH)_2SO_2 \xrightarrow[H_2O]{1.\ NaOCl, NaOH}$$

$$\begin{array}{c} NaO_3S—N—C_2H_5 \\ | \\ NH—C_2H_5 \end{array} \longrightarrow \begin{array}{c} HN—C_2H_5 \\ | \\ HN—C_2H_5 \end{array} \xrightarrow[H_2O, 25°]{NaOCl, NaOH} \begin{array}{c} N—C_2H_5 \\ \| \\ N—C_2H_5 \end{array}$$

Submitted by ROLAND OHME, HELMUT PREUSCHHOF, and
HANS-ULRICH HEYNE[1]
Checked by HARVEY W. TAYLOR and HENRY E. BAUMGARTEN

1. Procedure

Caution! Azoalkanes have been reported to have carcinogenic properties.[2,3] *Care should be taken to avoid inhalation of these substances and contact of them with the skin. It is advisable to prepare and handle these compounds in a good fume hood.*

A. *N,N'-Diethylsulfamide.* In a dry 2-l. three-necked round-bottomed flask fitted with a mechanical stirrer, a reflux condenser, a thermometer, and a dropping funnel, and protected from atmospheric moisture with calcium chloride-filled drying tubes, are placed 500 ml. of petroleum ether, 100 g. (2.20 moles) of ethylamine, and 140 g. (1.76 moles) of pyridine (Note 1). The stirred mixture is cooled in a dry ice-acetone bath to $-30°$ to $-15°$; then a solution of 120 g. (0.88 mole) of sulfuryl chloride in 220 ml. of petroleum ether is added, dropwise and with stirring, to the reaction flask at such a rate that the temperature remains below $-15°$. After addition is complete, the reaction mixture is stirred at room temperature for one hour. The petroleum ether layer is separated and discarded. The dark semisolid residue is made acidic by addition of aqueous $6M$ hydrochloric acid, and the acidic mixture is heated under reflux for 2 hours (Note 2). The resulting solution is extracted with ether in a continuous extractor (Note 3) until all of the diethylsulfamide has dissolved. The ether is evaporated using a rotary evaporator. The yield of crude N,N'-diethylsulfamide is 58–61 g. (44–45%), m.p. 65–67° (Note 4). This product is satisfactory for use in the following procedure without further purification.

B. *Azoethane*. In a 3-l. three-necked round-bottomed flask fitted with a mechanical stirrer, a reflux condenser, a thermometer, and a dropping funnel are placed 152 g. (1.00 mole) of N,N'-diethylsulfamide and 500 ml. (1.00 mole) of aqueous $2M$ sodium hydroxide. The sulfamide is brought into solution by warming the reaction flask. The reaction flask is cooled in a cold water bath and 715 ml. (1.00 mole) of aqueous $1.4M$ sodium hypochlorite (Note 5) is added dropwise with stirring. After addition is complete, the reaction mixture is stirred for 15 minutes at room temperature. The mixture is brought to pH 1 by addition of aqueous $6M$ hydrochloric acid and is stirred for an additional 30 minutes at 60° (Note 6). The mixture is cooled to room temperature and then is brought to pH 14 by addition of aqueous $2M$ sodium hydroxide (Note 7). Addition of 715 ml. (1.00 mole) of aqueous $1.4M$ sodium hypochlorite solution causes the separation of azoethane as a oil with a fruit-like odor. The mixture is extracted with three 100-ml. portions of toluene (Note 8). The toluene extracts are dried over anhydrous sodium sulfate and distilled through a 50-cm. packed column. The yield of azoethane, b.p. 58–59°, n^{20} D 1.3861, is 44–46 g. (51–54%) (Note 9).

2. Notes

1. The submitters dried the ethylamine and pyridine by distillation over potassium hydroxide pellets. The submitters used 600 ml. of petroleum ether, 113 g. (2.50 moles) of ethylamine, and 158 g. (2.00 moles) of pyridine to which was added 135 g. (1.00 mole) of sulfuryl chloride in 250 ml. of petroleum ether. In the United States ethylamine is sold in 100-g. quantities in sealed-glass vials (Eastman Organic Chemicals) or as the compressed gas in cylinders (Matheson Gas Products). The checkers used the contents of a freshly opened vial (without distillation) for each run as a matter of convenience. The checkers used either pentane or petroleum ether (b.p. 38–51°).

Note that step B requires the product from at least two [submitters' scale *and yield* (Note 4)] or three (checkers' scale) step A runs. Step B can be run at half scale with the same percentage yield.

2. The purpose of this step is to hydrolyze any alkyl imido compound that may have formed from the further reaction of the sulfamide.[4]

3. A convenient continuous extractor has been described earlier in this series.[5]

4. After purification by dissolving the crude product in ether and precipitating with petroleum ether, N,N'-diethylsulfamide is obtained as shiny white leaflets, m.p. 67°. The submitters reported a 54% yield of the purified product.

Under identical conditions from 78 g. (2.5 mole) of methylamine 71 g. (57%) of N,N'-dimethylsulfamide, m.p. 76°, may be obtained as fine white needles after recrystallization from benzene.

For sulfamides with larger alkyl groups (C_3 to C_6) the following procedure is preferred. To a stirred mixture of 135 g. (1.00 mole) of sulfuryl chloride and 500 ml. of chloroform is added, dropwise and with cooling to $-10°$ to $-5°$, a solution of 316 g. (4.00 moles) of pyridine in 400 ml. of chloroform followed by, with cooling to $-5°$ to $0°$, a solution of 2.5 moles of alkylamine in 600 ml. of chloroform. After addition is complete the mixture is stirred for 30 minutes at room temperature and then evaporated under reduced pressure to a thick brown liquid, to which aqueous $2M$ hydrochloric acid is added until the pyridine dissolves. On cooling of the acidic solution the crystalline sulfamide precipitates and is filtered. Any dissolved sulfamide may be recovered by extraction of the filtrate with ether. The crude product may be purified by recrystallization from 50% ethanol.

The pyridine used in the submitters' procedures apparently reacts with the sulfuryl chloride to form an intermediate quaternary pyridinium complex which undergoes aminolysis to yield the sulfamide.[6] However, in many instances the pyridine may be replaced by an equivalent quantity of the primary alkylamine being used.[4,7] Using this variation in the checkers, laboratory a 78% yield of N,N'-dicyclohexylsulfamide (compare with Table I) was obtained. Moreover, in the reaction of 4-aminospiro[cyclohexane-1,9'-fluorene] with sulfuryl chloride no sulfamide could be isolated from reactions run in the presence of pyridine (or triethylamine); however, a 54% (purified) yield of N,N'-dispiro[cyclohexane-1,9'-fluorene]-4-ylsulfamide was obtained when 2.7 *equivalents* of the amine (relative to sulfuryl chloride) were used. Probably the failure of the mixed pyridine-alkylamine technique was the result of combined bulk of the pyridinium complex and the amine.

5. The sodium hypochlorite solution was prepared by passing chlorine, with stirring and cooling, into 1.5 l. of aqueous 1.4M sodium hydroxide solution held at 0–5°.

In some small-scale preparations of this type in the checkers' laboratory, commercial household bleach (Chlorox®, 5.25% NaOCl) has been used and the course of the reaction has been followed by thin layer chromatography. The yields appear to be somewhat lower than those obtained with sodium hypochlorite prepared as described above. The obvious attractive alternative, preparation of potassium hypochlorite as described elsewhere in this series,[8] apparently has not been tried.

6. In the preparation of 2,2'-azoisobutane and azocyclohexane the acid hydrolysis step is not necessary and the two moles of sodium hypochlorite may be added in one step.

7. In the preparation of azomethane nitrogen is passed slowly through the reaction mixture using a gas-inlet tube during the second oxidation stage while the temperature is raised to 60°. The reflux condenser is fitted with a drying tube filled with potassium hydroxide pellets connected via rubber hose to two dry ice-cooled cold traps connected in series and terminated with a second drying tube filled with potassium hydroxide pellets. The azomethane collects in the cold traps. Redistillation gives a 39% yield of azomethane, b.p. 1°.

8. For the homologous azoalkanes ether, pentane, or petroleum ether may be used for extraction. The extraction solvent can be added before the addition of hypochlorite.[6]

9. The checkers used a 60-cm Vigreux column. Their product gave the following n.m.r. spectrum (CDCl$_3$): 3.77 (3H triplet, $J = 7$ Hz., CH$_3$), 1.17 p.p.m. (2H quartet, $J = 7$ Hz., CH $_2$).

3. Discussion

Azoalkanes have been prepared by oxidation of N,N'-dialkylhydrazines with copper(II) chloride[9] or with yellow mercury(II) oxide.[10,11] The dialkyl hydrazines are obtained by alkylation of N,N'-diformylhydrazine and subsequent hydrolysis,[9] by reduction of the corresponding azine with lithium aluminum hydride,[11] or by catalytic hydrogenation of the azine over a platinum catalyst.[10]

The present procedure may be used for the preparation of azoalkanes with alkyl, cycloalkyl, or aromatic substituents (Table I). Azo alkanes have been used as radical sources for inducing of radical reactions (*e.g.*, polymerization). The present procedure may also be used for the preparation of N,N'-dialkylhydrazines.[6] For this purpose only one equivalent of sodium hypochlorite solution is employed and the reaction mixture is worked up after its addition (yields: 60–95%).

TABLE I
PREPARATION OF AZOALKANES

R	RNHSO$_2$NHR		R—N=N—R	
	m.p.	Yield, %	b.p.	Yield, %
n-C$_3$H$_7$	118°	69	113–115°	54
n-C$_4$H$_9$	126°	66	59–60° (18 mm.)	54
t-C$_4$H$_9$	140–142°	68	109–110°	84
cyclo-C$_6$H$_{11}$	154°	59	m.p. 33–34°	80
p-NO$_2$-C$_6$H$_4$	197°	58	m.p. 216°	31

1. Institut für Organische Chemie, Deutsche Akademie der Wissenschaften, 1199 Berlin-Adlershof, D.D.R.
2. H. Druckrey, R. Preussmann, S. Ivanković, C. H. Schmidt, B. T. So, and C. Thomas, *Z. Krebsforsch.*, **67**, 31 (1965).
3. R. Preussmann, H. Druckrey, S. Ivankovic, and A. von Hondenberg, *Ann. N.Y. Acad. Sci.*, 163 (1969).
4. R. Sowada, *J. Prakt. Chem.*, (4) **20**, 310 (1963).
5. See Note 10 in G. Billek, *Org. Syn.*, **43**, 52 (1963).

6. R. Ohme and H. Preuschhof, *Justus Liebigs Ann. Chem.*, **713**, 74 (1968).
7. J. C. Stowell, *J. Org. Chem.*, **32**, 2360 (1967).
8. M. S. Newman and H. L. Holmes, *Org. Syn.*, Coll. Vol. **2**, 428 (1943).
9. J. L. Weininger and O. K. Rice, *J. Amer. Chem. Soc.*, **74**, 6216 (1952).
10. A. U. Blackham and N. L. Eatough, *J. Amer. Chem. Soc.*, **84**, 2922 (1962).
11. R. Renaud and L. C. Leitch, *Can. J. Chem.*, **32**, 545 (1954).

BENZYL CHLOROMETHYL ETHER

(Ether, benzyl chloromethyl)

$$C_6H_5CH_2OH + HCl + HCHO \xrightarrow{10°} C_6H_5CH_2OCH_2Cl$$

Submitted by D. S. CONNOR, G. W. KLEIN, and G. N. TAYLOR[1]
Checked by R. KEESE, M. GOLDBERG, and A. ESCHENMOSER

1. Procedure

Caution! This procedure should be performed in a hood to avoid exposure to hydrogen chloride.

To a 500-ml. three-necked flask equipped with a Trubore stirrer with a Teflon paddle, a gas inlet tube, and a $-10°$ to $100°$ thermometer is added 150 g. (1.39 moles) of benzyl alcohol (Note 1) and 120 g. (1.48 moles) of an aqueous 37% formaldehyde solution (Note 2). The solution is cooled to 5° in an ice-salt bath and anhydrous hydrogen chloride is bubbled into the solution with vigorous stirring. The temperature is maintained below 10° (Note 3) during the 6–8 hours required to saturate the solution with hydrogen chloride. After saturation is complete, hydrogen chloride is passed through the solution for an additional hour at 10° and the mixture is then transferred to a separatory funnel. The reaction mixture separates into two layers, the lower of which is discarded. The upper layer is dried over anhydrous calcium chloride and then placed on a rotary evaporator at aspirator pressure for one hour to remove hydrogen chloride. The undistilled benzyl chloromethyl ether (Note 4) weighs 192–212 g. (88–97%) (Note 5). Purification by distillation under reduced pressure (Note 6) may be unnecessary. Distillation to obtain the fraction, b.p. 53–56° (1.5 mm.)

or 96–99° (11 mm.) separates 141–171 g. (65–80%) of benzyl chloromethyl ether, n^{20} D 1.5268–1.5279 (Note 7).

2. Notes

1. Chlorine free benzyl alcohol is used.
2. s-Trioxane (0.493 mole) may be substituted for aqueous 37% formaldehyde solution with no change in the procedure.
3. Temperatures in excess of 10° lead to the formation of substantial amounts of dibenzylformal.
4. Caution must be used in handling benzyl chloromethyl ether since it is a mild lachrymator and reacts with water to form hydrogen chloride.
5. The crude product exhibits n.m.r. singlets (CCl_4 solution) at 7.29(5H), 5.41(2H), and 4.68 p.p.m.(2H). Both n.m.r. and gas chromatographic analyses indicate a purity of greater than 90% with the major, and in most cases only, impurity being dibenzylformal. Gas chromatographic analysis was obtained at 155° with a 2 m. × 0.7 cm. column packed with silicone fluid, No. 710, suspended on 60–80 mesh firebrick. The checkers have found benzyl chloride to be the main contaminant of the undistilled product (n.m.r. analysis).
6. Complete decomposition occurs if distillation is attempted at atmospheric pressure.
7. Small samples may readily be distilled with little decomposition, while large samples decompose significantly during distillation. The undistilled material may be satisfactory for use in alkylation reactions without purification. The checkers have found that 200 g. quantities of the product can reproducibly be distilled at approximately 1 mm. The reported physical constants for benzyl chloromethyl ether are: b.p. 96–98° (9.5 mm.)[2], n^{20} D 1.5264–1.5292.[7,12]

3. Discussion

Benzyl chloromethyl ether is useful for the introduction of a potential hydroxymethyl group in alkylation reactions. Hill and Keach[3] were the first to use this method and found it convenient in barbiturate syntheses. Graham and McQuillin,[4] and Graham, McQuillin, and Simpson,[5] have extended the scope of the alkylation reaction to various ketone derivatives. They have also investigated the conditions for obtaining maximum C-alkylation and the stereochemistry of alkylation in various octalone systems.[4] Alkylation of ketones followed by sodium borohydride reduction and catalytic hydrogenolysis represents a convenient method for obtaining 1,3-diols.[4] Similarly, a Wolff-Kishner reduction and catalytic hydrogenolysis gives a primary alcohol.[4] A procedure of this type has been used for obtaining bridgehead methanol derivatives of bicyclic compounds.[6]

Several other alkylation reactions of benzyl chloromethyl ether have been reported using phosphorus compounds as nucleophiles.[7] Hydrolysis and alcoholysis reactions of the reagent[8] have been investigated along with the addition of the chloroether to propylene in the presence of zinc chloride.[9] The alkylation of enamines with benzyl bromomethyl ether has been reported.[10]

Benzyl chloromethyl ether has been prepared from benzyl alcohol, aqueous formaldehyde solution, and hydrogen chloride.[3,11,12] Gaseous formaldehyde[14] and trioxane[13] have also been used. This chloromethyl ether has also been prepared by the chlorination of benzyl methyl ether.[9] The present procedure is based on the first method, but avoids the use of a large excess of formaldehyde and provides a considerably simplified isolation method.

1. Department of Chemistry, Yale University, New Haven, Connecticut 06520.
2. A. Rieche and H. Gross, *Chem. Ber.*, **93**, 259 (1960).
3. A. J. Hill and D. T. Keach, *J. Amer. Chem. Soc.*, **48**, 257 (1926).
4. C. L. Graham and F. J. McQuillin, *J. Chem. Soc.*, 4634 (1963).
5. C. L. Graham, F. J. McQuillin, and P. L. Simpson, *Proc. Chem. Soc. (London)*, 136 (1963).
6. K. B. Wiberg and G. W. Klein, *Tetrahedron Lett.*, 1043 (1963).
7. V. S. Abramov, E. V. Sergeeva and I. V. Chelpanova, *J. Gen. Chem. USSR*, **14**, 1030 (1944), [*C. A.*, **41**, 700 (1947)].
8. H. Böhme and A. Dörries, *Chem. Ber.*, **89**, 719 (1956).
9. H. Böhme and A. Dörries, *Chem. Ber.*, **89**, 723 (1956).
10. A. T. Blomquist and E. J. Moriconi, *J. Org. Chem.*, **26**, 3761 (1961).

11. P. Carré, *Compt. Rend.*, **186**, 1629 (1928); *Bull. Soc. Chim. Fr.*, **43**, 767 (1928).
12. Sh. Mamedov, M. A. Avanesyan, and B. M. Alieva, *Zh. Obshch. Khim.*, **32**, 3635 (1962); [*C. A.*, **58**, 12444 (1963)].
13. S. Sabetay and P. Schving, *Bull. Soc. Chim. Fr.*, **43**, 1341 (1928).

3-BUTYL-2-METHYLHEPT-1-EN-3-OL

$$n\text{-}C_4H_9Br + 2\ Li \rightarrow n\text{-}C_4H_9Li + LiBr$$

$$2\ n\text{-}C_4H_9Li + CH_2=\underset{CH_3}{C}-\overset{O}{\overset{\|}{C}}OCH_3 \rightarrow CH_2=\underset{CH_3}{\overset{}{C}}-\underset{n\text{-}C_4H_9}{\overset{n\text{-}C_4H_9}{C}}-OLi + CH_3OLi$$

$$CH_2=\underset{CH_3}{\underset{|}{\overset{n\text{-}C_4H_9}{\overset{|}{C}}}}-\underset{n\text{-}C_4H_9}{\overset{}{C}}-OLi \xrightarrow[H_2O]{HCl} CH_2=\underset{CH_3}{\underset{|}{\overset{n\text{-}C_4H_9}{\overset{|}{C}}}}-\underset{n\text{-}C_4H_9}{\overset{}{C}}-OH + LiCl$$

Submitted by P. J. PEARCE,[1] D. H. RICHARDS, and N. F. SCILLY
Checked by N. COHEN, R. LOPRESTI, and A. BROSSI

1. Procedure

A 2-l. four-necked flask equipped with a sealed, Teflon-paddle stirrer, a mercury thermometer, a gas inlet tube, and a dropping funnel is charged with 1.2 l. of anhydrous tetrahydrofuran (Note 1) and 50 g. (7.1-g. atoms) of lithium pieces (Note 2) under an atmosphere of prepurified nitrogen. The stirred mixture is cooled to $-20°$ by means of a dry ice-acetone bath and a mixture of 100 g. (1.00 mole) of methyl methacrylate (Note 3), and 411 g. (3.0 moles) of *n*-butyl bromide (Note 4) is added dropwise over a period of 3–4 hours. During this addition, an exothermic reaction ensues which is controlled at $-20°$ (Note 5), and on completion of the addition, the vessel is maintained at this temperature, with stirring, for an additional 30 minutes. The contents of the flask are then filtered with suction through a

70-mm.-diameter, slit-sieve Buchner funnel, without filter paper, to remove the excess lithium metal. The filtrate is concentrated on a rotary evaporator at water aspirator pressure. The residual lithium alcoholate is hydrolyzed by the addition of 1 l. of aqueous 10% hydrochloric acid, with external cooling by means of an ice bath. The liberated alcohol is extracted with two 400-ml. portions of ether and the combined ether extracts are washed with two 400-ml. portions of water and then dried over 100 g. of anhydrous magnesium sulphate. After suction filtration and removal of the ether on a rotary evaporator at aspirator pressure, the crude alcohol is distilled under reduced pressure through a 40-cm. Vigreux column to separate 147–158 g. (80–86%) of 3-butyl-2-methylhept-1-en-3-ol, b.p. 80° (1mm.).

The purity of the product, determined by gas chromatographic analysis, is greater than 99%.

2. Notes

1. Laboratory reagent grade (stabilized) tetrahydrofuran was allowed to stand over molecular sieves for 24 hours, refluxed for 2 hours with sodium wire, and finally distilled and used within 48 hours. The checkers found that it was convenient simply to percolate the tetrahydrofuran, after preliminary drying over molecular sieves, through a column of grade I, neutral aluminum oxide, under nitrogen, directly into the reaction flask, until the required volume of solvent was collected.

2. A convenient form of lithium metal can be purchased from Associated Lead Manufacturers Ltd., 14 Gresham Street, London. A typical analysis shows a purity of 99.6% and it can be obtained as 1.3-cm.-diameter rod coated with petroleum jelly. A comparable form of lithium metal can be purchased from Ventron Corporation, Chemicals Division, Beverly, Massachusetts. Preparation for use involves weighing, washing with petroleum ether (b.p. 40–60°), and cutting the rod by scissors so that the pieces fall into the reaction vessel. The rod is cut into pieces about 0.5 cm. long that have an average weight of 0.3 g. per piece. Since excess lithium is employed in this reaction, accurate weighing is unnecessary.

3. Laboratory reagent grade methyl methacrylate monomer was dried over powdered calcium hydride and freshly distilled before use. The checkers found that identical yields could be obtained when Matheson Coleman and Bell Chromatoquality methyl methacrylate monomer was used as received with no purification.

4. Laboratory reagent grade n-butyl bromide (greater than 98% pure) was used after drying over molecular sieves.

5. The reaction is highly exothermic and the submitters have found that isothermal conditions are best maintained by using cooling equipment consisting of a cooling bath seated on a pneumatically operated labjack and controlled by a temperature sensor which is attached to the thermometer dipping into the reaction vessel. This equipment, known as Jack-o-matic, is supplied by Instruments for Research and Industry, Cheltenham, Pennsylvania.

3. Discussion

This method is of quite general applicability and the carbonyl compound may be an aldehyde, a ketone, or an ester.[2] Similarly, the halide may be chloride, bromide, or iodide although yields are generally lower with iodides. Alkyl and aryl halides react with equal facility and the alkyl halide may be primary, secondary, or tertiary. A few examples of the yields obtained with a variety of reagents are given in Table I (the yields quoted are obtained by g.l.c. analysis of the reaction mixture using an internal standard).

For maximum yield, care must be taken to ensure that the rate of addition of the reagents is not excessive. If this occurs then the alkyllithium is generated in the presence of significant amounts of unchanged alkyl halide and the Wurtz condensation reaction may be favored. The rate of formation of the alkyllithium is proportional to the surface area of the lithium metal and so at a constant rate of addition the Wurtz condensation is reduced by an increase in the lithium surface available for reaction.

Excess alkyl halide is required to compensate for these side reactions, although commonly only 10–20% excess is used

TABLE I

Carbonyl Compound	Halide	Product	Yield, %
Propionaldehyde	Ethyl bromide	3-Pentanol	90
Benzaldehyde	Chlorobenzene	Benzhydrol	100
Di-n-butyl ketone	n-Butyl bromide	Tri-n-butyl carbinol	91
Ethyl formate	n-Butyl bromide	Nonan-5-ol	91
Acrolein	Ethyl bromide	Pent-1-en-3-ol	90
Butyraldehyde	sec-Butyl bromide	3-Methylheptan-4-ol	89

rather than the 50% quoted in the method above. The yields given in the table are those obtained with 20% excess halide. The submitters have scaled up the reaction by a factor of 40 with no lowering of yield.

The technique is more efficient than the conventional Grignard reaction for three main reasons: (1) it is a one-stage process; (2) the yields are generally higher; and (3) the final product isolation is cleaner and more convenient.

1. Explosives Research and Development Establishment, Ministry of Defense, Waltham Abbey, Essex, U.K.
2. P. J. Pearce, D. H. Richards, and N. F. Scilly, *Chem. Commun.*, 1160 (1970); British Patent Application No. 61956 (1969).

CONTROLLED-POTENTIAL ELECTROLYTIC REDUCTION: 1,1-BIS(BROMOMETHYL)CYCLOPROPANE

$$C(CH_2Br)_4 + 2e^- \xrightarrow[\substack{(n\text{-}C_4H_9)_4NBr,\\(CH_3)_2NCHO}]{\substack{\text{Hg cathode}\\(-1.8 \text{ volt vs. sce})}} \triangleright(CH_2Br)_2 + 2Br^-$$

Submitted by M. R. Rifi[1]
Checked by Edith Feng and Herbert O. House

1. Procedure

Caution! Since bromine is liberated during the electrolysis, the reaction should be conducted in a hood.

The electrolysis cell (Note 1) is assembled in a 1000-ml. flat-bottomed Pyrex reaction kettle. The Pyrex cover contains four standard taper outer joints in which are mounted: (1) a 11.5-cm. length of 2-cm.-diameter carbon rod (Note 2) surrounded by a 15 × 5.5-cm.-diameter porous porcelain cup (Note 3); (2) a 31-cm. × 8-mm.-diameter length of soda-lime glass tubing (Note 4) with a short length of 0.6-mm.-diameter platinum wire fused into the bottom: (3) a tee-tube fitted to hold a thermometer and to allow nitrogen to be passed into the reaction vessel: and (4) a saturated calomel reference electrode fitted with successive salt bridges containing aqueous $1M$ sodium nitrate and $1.5M$ tetraethylammonium tetrafluoroborate in dimethylformamide (Note 5). The cover is also equipped with a suitable clamp so that it may be fastened to the reaction kettle during the electrolysis.

A sufficient quantity of mercury (about 700 g.) is added to the reaction kettle to form a cathode pool 1-cm. deep. A Tefloncovered magnetic stirring bar is placed on this mercury pool. Then a solution of 25.0 g. (0.065 mole) of pentaerythrityl tetrabromide (Note 6) in 250 ml. of $0.2M$ tetra-n-butylammonium bromide (Note 7) in N,N-dimethylformamide (Note 8) is added to the reaction vessel. An additional 175 ml. (Note 9) of $0.2M$ tetra-n-butylammonium bromide (Note 7) in N,N-dimethylformamide (Note 8) is added to the porous porcelain cup surrounding the carbon-rod anode and the cover is clamped to the reaction kettle. The glass tubing with a platinum wire contact sealed in the bottom is adjusted so that all of the exposed platinum is below the mercury surface (Note 10). The bottom of the porous porcelain cup should be a sufficient distance above the mercury pool so as not to interfere with the magnetic stirring bar. The salt bridge associated with the calomel reference electrode (Note 5) should be adjusted so that the lower porous Vycor plug is between 5 and 10 mm. above the surface of the mercury pool. Before the electrolysis is begun a 0.5-ml. aliquot of the reaction solution should be removed from the cathode compartment and analyzed polarographically (Note 11), and the electrical resistances between the cathode and anode and between the cathode and the reference electrode should be measured with a suitable resistance bridge (Note 12). If all electrical connections are satisfactory the cathode-anode resistance should be in the range 20–30 ohms and the cathode-reference resistance should be in the range 5000–15,000 ohms. The electrical leads from the anode and cathode should be connected with a direct current source whose potential may be conveniently adjusted from 0 to 40 volts with a continuous current output of at least 1 amp. A suitable d.c. voltmeter is mounted in parallel with the anode and cathode leads and a d.c. ammeter, capable of measuring currents of 0 to 3 amps is placed in series in either of the two leads. Finally, a vacuum-tube d.c. voltmeter or some equivalent *high-impedance* (Note 13) potential measuring device is connected to measure the potential difference between the cathode and the reference electrode. The reaction kettle should be placed in a nonmagnetic bath to which cooling water may be added if necessary and magnetic stirring started. A very slow stream of nitrogen (1–2 ml. per minute) is passed through the apparatus throughout the electrolysis (Note 14). The potential of the direct current source is adjusted to give a potential difference of 1.7 to 1.8 volts between the cathode and the reference electrode (Note 15) and the current source is adjusted at 10–15 minute intervals to maintain this potential difference throughout the electrolysis. It is convenient to keep a record of time and the current passing through the cell so that the time when the theoretical amount of electricity (12,500 coulombs or 3.5 amp.-hours) has been passed through the cell can be estimated (Note 16). During the electrolysis the temperature of the catholyte solution is kept below 40° (Note 17) by the use of external cooling if necessary. When approximately the theoretical amount of electricity has been passed through the electrolysis cell (Note 16, typically 4–6 hours), a 0.5-ml. aliquot is removed and analyzed polarographically (Note 11). The electrolysis is continued until the polarographic analysis (Note 11) indicates the consumption of practically all the pentaerythrityl tetrabromide.

The solutions are removed from the cathode and anode (Note 18) compartments and added to 200 ml. of water. The resulting mixture is extracted with four 150-ml. portions of pentane and the combined pentane extracts are washed with water, dried over anhydrous sodium sulfate, and then concentrated by distillation through a short Vigreux column (Note 19). The remaining pentane is removed by distillation and the residual yellow liquid is distilled under reduced pressure. The 1,1-bis-(bromomethyl)cyclopropane is collected as 6.9–8.5 g. (47–58%) of colorless liquid boiling at 65–67° (5 mm.), n^{25} D 1.5341–1.5347 (Note 20).

2. Notes

1. The electrolysis cell designed by the checkers is shown in Figure 1. The adapters, which hold the reference electrode salt bridge (Note 5) and the glass tube with the platinum contact and the tee tube for the nitrogen inlet, and the thermometer are all commercially available from Ace Glass, Inc., Vineland, New Jersey. The adapter which supports the carbon rod anode and the surrounding porous porcelain cup was machined from a Teflon rod; the dimensions of the adapter used by the checkers are indicated in Figure 2. Holes were drilled in the porcelain cup to permit it to be fastened to the Teflon adapter with three stainless steel machine screws. Electrical contact between the carbon anode and the wire to the external circuitry was achieved by drilling and tapping the end of the carbon rod for a small machine screw. The arrangement used for the remainder of the electrical circuit is indicated in Figure 1.

2. Carbon rods of approximately the dimensions indicated are commercially available from welding supply companies.

3. A suitable porous porcelain cup (Coors 700, unglazed) may be purchased from the Arthur H. Thomas Company.

4. To obtain a satisfactory seal between the platinum and the glass, soda-lime glass rather than Pyrex glass should be employed. Before the platinum wire is sealed into the glass a length of copper wire should be silver soldered to the platinum to provide an accessible electrical lead at the top of the glass tubing.

5. Any commercial saturated calomel electrode of convenient

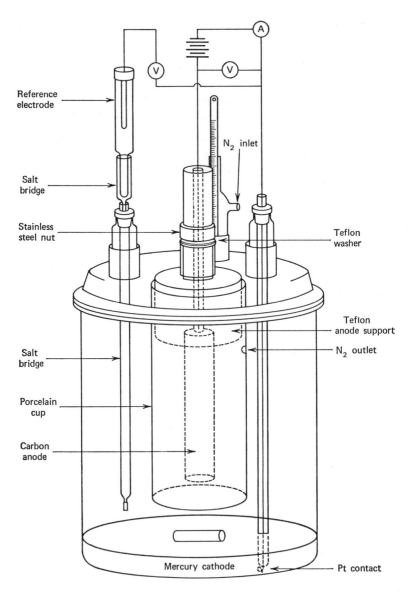

Figure 1. Cell for controlled-potential electrolytic reduction.

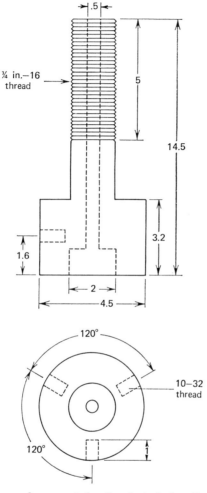

Figure 2. Teflon anode support for the electrolysis cell. Unless otherwise stated, the dimensions are in centimeters.

dimensions may be employed. The arrangement of the salt bridges between the calomel electrode and the reaction solution is illustrated in Figure 3. The Teflon tubing is available from Bolab, Inc., 359 Main Street, Reading, Massachusetts 01867, and the porous Vycor plugs are cut from lengths of $\frac{1}{8}$-in.-diameter porous Vycor rod ("thirsty" glass or Corning Vycor No. 7930) available from the Electronic Parts Department,

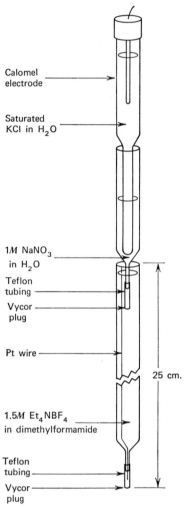

Figure 3. Reference electrode and salt bridges for controlled potential electrolysis.

Corning Glass Works, Houghton Park, Corning, New York, 14830. The intermediate salt bridge containing aqueous sodium nitrate is used to prevent precipitation of the insoluble potassium tetrafluoroborate at the small fiber in the tip of the calomel electrode.[2] To minimize resistance in the long, lower nonaqueous salt bridge, a concentrated (1.5M) solution of tetraethylammonium tetrafluoroborate[3] in N,N-dimethylformamide is used with a length of platinum wire inside the tube. Tetraethylammonium tetrafluoroborate was prepared by mixing 5.3 g. (0.025 mole) of tetraethylammonium bromide (Eastman Organic Chemicals) and 3.6 ml. (*ca.* 0.026 mole) of aqueous 48–50% fluoroboric acid (Allied Chemical Corporation) in 8 ml. of water. The resulting mixture was concentrated under reduced pressure, diluted with ether and filtered to separate 4.6 g. (85%) of the crude tetrafluoroborate salt, m.p. 375–378° dec. Two recrystallizations from methanol-petroleum ether mixtures afforded 3.7 g. (69%) of the pure tetraethylammonium tetrafluoroborate as white needles, m.p. 377–378° dec. (after drying).

6. Crude pentaerythrityl tetrabromide, purchased from Columbia Organic Chemicals Company, Inc., was recrystallized from chloroform to separate the tetrabromide as tan needles, m.p. 158–160°. Alternatively, this material may be obtained by the procedure described in a previous volume of this series.[4]

7. Tetra-*n*-butylammonium bromide, obtained from Eastman Organic Chemicals, was recrystallized from chloroform. The white prisms that separated were pulverized and then dried under reduced pressure to give the pure salt, m.p. 116–117.5°.

8. N,N-Dimethylformamide, obtained from Allied Chemical Corporation, was purified by distillation under reduced pressure, b.p. 39–41° (6 mm.).

9. Since the amount of solvent in the anode compartment is slowly depleted as the electrolysis proceeds, it is convenient to begin the electrolysis with the level of solution in the anode compartment about 9 cm. above the level of the solution in the cathode compartment. Alternatively, additional $0.2M$ tetra-*n*-butylammonium bromide in N,N-dimethylformamide may be added to the anode compartment as the electrolysis proceeds.

10. To avoid competing liberation of hydrogen at the cathode, no platinum should be exposed to the catholyte solution.

11. The polarographic analysis is obtained at a dropping mercury electrode with any conventional three-electrode polarograph employing a saturated calomel electrode as the reference. The checkers added the 0.5-ml. aliquots of the

reaction mixture to 10-ml. portions of $0.2M$ tetra-n-butylammonium bromide (Note 7) in N,N-dimethylformamide (Note 8). The half-wave potentials ($E_{1/2}$ vs. sce) for the reduction of pentaerythrityl tetrabromide and 1,1-bis-(bromomethyl)-cyclopropane are -1.71 ($\alpha n = 0.44$) volts and -2.18 ($\alpha n = 0.31$) volts, respectively.

12. The checkers measured these resistances with a Serfass Conductivity Bridge, Model RCM 15, employing a 1000-Hz. alternating current.

13. Since the resistance in the circuit containing the reference electrode is approximately 10,000 ohms, an accurate measure of the cathode-reference electrode potential can only be obtained by the use of a potential measuring device with an imput impedance of at least 100,000 ohms. Although a vacuum-tube voltmeter (VTVM, typical imput impedance 11×10^6 ohms) is suitable for this purpose, a common multimeter (VOM, typical imput impedance 20,000 ohms per volt) *is not a satisfactory alternative*.

14. This slow stream of nitrogen passes through the hole about 2.5 cm. from the top of the porous cup surrounding the anode compartment and sweeps the bromine formed at the anode out the top of the apparatus through the hole in the center of the Teflon adapter.

15. If the cathode-reference potential is allowed to rise significantly above -1.8 volts, further reduction of the 1,1-bis-(bromomethyl)cyclopropane will occur. The submitter reports that the dibromide may be reduced to spiropentane in 39% yield by employing a cathode-reference potential of -2.0 volts to -2.2 volts.

16. The checkers found that about 1.3 times the theoretical amount of electricity was passed through the cell during the time required to consume the starting material. Presumably the excess electricity is consumed partially in the reduction of impurities (*e.g.*, oxygen) and partially in the further reduction of the dibromide to spiropentane (Note 15).

17. This temperature limit was selected to avoid the possible loss of solvent and the volatile product in the slow stream of

nitrogen gas being passed through the cell. Since the rate of the electrolytic reduction is increased with an increase in the reaction temperature, it is advantageous to maintain the temperature of the reaction solution in the range 35–40°.

18. Since some diffusion of the product from the cathode compartment to the anode compartment is probable, both solutions are subjected to the isolation procedure.

19. To avoid the possible loss of the volatile product, the pentane solvent should be removed by distillation through a 20–30-cm. Vigreux column rather than by distillation under reduced pressure in a rotary evaporator.

20. On a gas chromatography column, packed with Apiezon M suspended on Chromosorb P and heated to 70°, the product exhibits a single peak with a retention time of 23.2 minutes. The sample exhibits infrared absorption (CCl_4 solution) at 3100, 3030, 2985, 1440, 1340, and 1235 cm.$^{-1}$ with n.m.r. singlets (CCl_4 solution) at 3.45 (CH_2Br) and 0.90 p.p.m. (cyclopropyl CH_2). The mass spectrum of the sample has abundant fragment peaks at m/e 149, 147, 67, 41, and 39.

3. Discussion

1,1-bis-(Bromomethyl)cyclopropane has been obtained as one component in a mixture of isomeric bromides by reaction of methylenecyclobutane with bromine[5] and by reaction of 1,1-bis-(hydroxymethyl)cyclopropane with phosphorus tribromide.[6-8] The present method illustrates a general procedure for the preparation of cyclopropane and cyclobutane derivatives by the electrolytic reduction of 1,3-and 1,4-dihalides.[9-13] In at least some cases, this method is clearly superior to the reductive cyclizations of dihalides effected with metals or with chromium(II) salts. Examples of this reductive ring closure are provided in Table I. The reaction appears to give the best results when dibromides rather than dichlorides are used as starting materials and when an aprotic solvent such as N,N-dimethylformamide or acetonitrile is used. These electrolytic ring closures proceed by way of a stepwise reaction mechanism.[13]

TABLE I
ELECTROLYTIC REDUCTIVE RING CLOSURE OF DIHALIDES

Dihalide	Product(s)	Yield, %	Ref.
$Br(CH_2)_3Br$	△ (only product)	60–80	9, 12
$Br(CH_2)_4Br$	25% □ + 75% $CH_3CH_2CH_2CH_3$	—	9
$Br(CH_2)_5CH_3$	$CH_3(CH_2)_3CH_3$ (only product)	80	9
$C_6H_5CHCH_2CH_2Br$ $\vert$ Br	C_6H_5—△	70	11
▷$(CH_2Br)_2$	▷◁	39	12
CH_3 CH_3 ╲◇╱ Br Br	CH_3—◇—CH_3	55–94	9
Cl—◇—Br	60% ◇ + 20% □ + 10% □	—	9
$CH_3CHCH_2CHCH_3$ $\vert$ $\vert$ Br Br (either meso or racemic isomer)	CH_3—△—CH_3 + (both cis and trans isomers) n-C_5H_{12} + isomeric pentenes	—	13

1. Chemicals and Plastics Division, Union Carbide Corporation, Bound Brook, New Jersey 08805.
2. B. McDuffie, L. B. Anderson, and C. N. Reilley, *Anal. Chem.*, **38**, 883 (1966).
3. (a) H. O. House, E. Feng, and N. P. Peet, *J. Org. Chem.*, **36**, 2371 (1971); (b) C. W. Wheeler, Jr., and R. A. Sandstedt, *J. Amer. Chem. Soc.*, **77**, 2025 (1955).

4. H. L. Herzog, *Org. Syn.*, Coll. Vol. **4**, 753 (1963).
5. D. E. Applequist and J. D. Roberts, *J. Amer. Chem. Soc.*, **78**, 874 (1956).
6. Ya. M. Slobodin and I. N. Shokhor, *J. Gen. Chem. USSR*, **21**, 2231 (1951).
7. N. Zelinsky, *Ber.*, **46**, 160 (1913).
8. W. M. Schubert and S. M. Leahy, Jr., *J. Amer. Chem. Soc.*, **79**, 381 (1957).
9. M. R. Rifi, *J. Amer. Chem. Soc.*, **89**, 4442 (1967).
10. M. R. Rifi, *Tetrahedron Lett.*, 1043 (1969).
11. R. Gerdil, *Helv. Chim. Acta*, **53**, 2100 (1970).
12. M. R. Rifi, unpublished work.
13. A. J. Fry and W. E. Britton, *Tetrahedron Lett.*, 4363 (1971).

1,2-DIAROYLCYCLOPROPANES: *trans*-1,2-DIBENZOYLCYCLOPROPANE

(Cyclopropane, *trans*-1,2-dibenzoyl)

$$C_6H_5COCH_2CH_2CH_2COC_6H_5 + I_2 + 2\,NaOH \xrightarrow[25-40°]{CH_3OH}$$

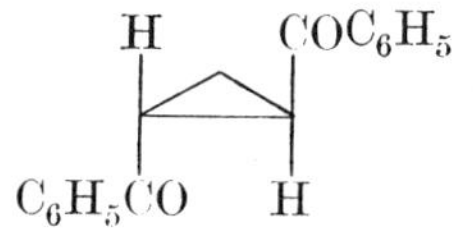

$$\underset{C_6H_5CO\quad H}{\overset{H\quad COC_6H_5}{\triangle}} + 2\,NaI + 2\,H_2O$$

Submitted by ISMAEL COLON, GARY W. GRIFFIN, and E. J. O'CONNELL, JR.[1]
Checked by STEPHEN P. PETERS and KENNETH B. WIBERG

1. Procedure

To a 1-l. three-necked, round-bottomed flask equipped with a magnetic stirrer and a dropping funnel are added 35 g. (0.14 mole) of 1,3-dibenzoylpropane (Note 1) and a solution of 11.2 g. (0.28 mole) of sodium hydroxide in 400 ml. of methanol. The mixture is warmed to 45° with stirring to dissolve the diketone. It is allowed to cool to 40°, and a solution of 35 g. (0.14 mole) of iodine in 200 ml. of methanol is then slowly added to the stirred solution from the dropping funnel (Note 2). After the addition of iodine is completed, the resulting clear solution is stirred at room temperature for 1.5 hours. During this period, a white solid precipitates (Note 3). The mixture is filtered and the filtrate is placed in a round-bottomed flask. The white solid is washed with four 100-ml. portions of water and dried at a pressure of 1 mm. at 57° for 15 hours to leave 23–25 g. (66–72%) of trans-1,2-dibenzoylcyclopropane, m.p. 103–104° (Note 4). Additional material is obtained on evaporation of the filtrate using a rotary evaporator. The pale red solid residue is treated with 25 ml. of aqueous 10% sodium bisulfite (Note 5), filtered, and dried as above giving 8–9 g. (23–26%) of the product, m.p. 95–98°. Recrystallization from methanol affords greater than 90% recovery of trans-1,2-dibenzoylcyclopropane, m.p. 102–103°. For the recrystallization of a 10-g. sample 200 ml. of methanol is used.

2. Notes

1. 1,3-Dibenzoylpropane was prepared by the method described for 1,4-dibenzoylbutane in *Organic Syntheses*.[3]

2. The iodine solution is added at a rate such that the color of the iodine is continually discharged by the rapidly stirred solution.

3. In some cases the precipitate may begin to form during the addition of the iodine solution. This has no effect on the yield.

4. Pertinent spectral data include a carbonyl stretching band in the infrared at 1665 cm.$^{-1}$ and ultraviolet absorption maxima at 245 mμ (ϵ 31,300), 278 mμ (ϵ 2130), and 317 mμ (ϵ 194). The n.m.r. spectrum (CCl$_4$ solution) has two triplets of equal intensity at 3.32 and 1.68 p.p.m. in addition to the aromatic proton bands at 7.3–8.2 p.p.m. The integrated peak areas are in the ratio 1:1:5.

5. Sodium bisulfite solution is added to reduce unchanged iodine.

3. Discussion

1,3-Dibenzoylcyclopropane has been prepared by the method described here[4] and previously by Conant and Lutz.[2] Both procedures use 1,3-dibenzoylpropane as the reactant. The present procedure is accomplished in one step under very mild conditions and in nearly quantitative yield. The method of

Conant and Lutz is a two-step process involving initial dibromination of the diketone followed by ring closure with zinc and sodium iodide. The overall yield of *trans*-1,2-dibenzoylcyclopropane is approximately 15%. The submitters have extended the described method to the preparation of other *trans*-1,2-diaroylcyclopropanes, namely *trans*-1,2-di-*p*-methoxybenzoylcyclopropane and *trans*-1,2-di(2-methoxy-5-methylbenzoyl)cyclopropane. In theory the method is applicable to closure of all α,α'-disubstituted propanes having acidic hydrogens on both α-carbons. A probable mechanism for this transformation is as follows:

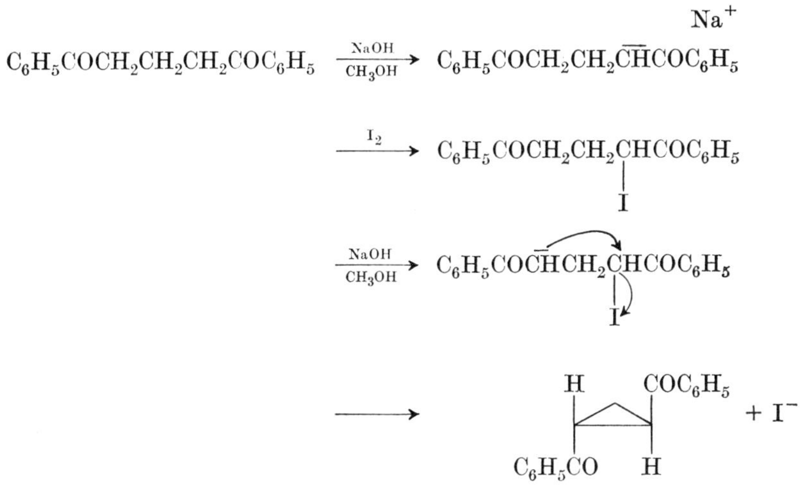

$$C_6H_5COCH_2CH_2CH_2COC_6H_5 \xrightarrow[CH_3OH]{NaOH} C_6H_5COCH_2CH_2\overline{CH}COC_6H_5 \ Na^+$$

$$\xrightarrow{I_2} C_6H_5COCH_2CH_2\underset{I}{CHCOC_6H_5}$$

$$\xrightarrow[CH_3OH]{NaOH} C_6H_5CO\overline{CH}CH_2\underset{I}{CHCOC_6H_5}$$

$$\longrightarrow \underset{C_6H_5CO\ \ \ \ H}{\overset{H\ \ \ \ \ COC_6H_5}{\triangle}} + I^-$$

1. Department of Chemistry, Fairfield University, Fairfield, Connecticut 06430.
2. J. B. Conant and R. E. Lutz, *J. Amer. Chem. Soc.*, **49**, 1083 (1927).
3. R. C. Fuson and J. T. Walker, *Org. Syn.*, Coll. Vol. **2**, 169 (1943).
4. G. W. Griffin, E. J. O'Connell, and H. A. Hammond, *J. Amer. Chem. Soc.*, **83**, 1003 (1963).

DIPHENYLKETENE

(Ketene, diphenyl)

$$(C_6H_5)_2CHCO_2H \xrightarrow[C_6H_6,\ \text{reflux}]{SOCl_2} (C_6H_5)_2CHCOCl + SO_2 + HCl$$

$$(C_6H_5)_2CHCOCl \xrightarrow[(C_2H_5)_2O,\ 0°]{(C_2H_5)_3N} (C_6H_5)_2C=C=O + (C_2H_5)_3N \cdot HCl$$

Submitted by EDWARD C. TAYLOR, ALEXANDER McKILLOP, and GEORGE H. HAWKS[1]
Checked by C. J. MICHEJDA, D. D. VON RIESEN, R. W. COMNICK, and HENRY E. BAUMGARTEN

1. Procedure

A. *Diphenylacetyl Chloride.* In a 500-ml. three-necked flask equipped with a dropping funnel and a reflux condenser carrying a calcium chloride drying tube are placed 50.0 g. (0.23 mole) of diphenylacetic acid (Note 1) and 150 ml. of thiophene-free anhydrous benzene. The mixture is heated under reflux and 132 g. (80 ml., 1.11 mole) of thionyl chloride is added dropwise during 30 minutes. Refluxing is continued for 7 hours more; then the benzene and excess thionyl chloride are removed by distillation under reduced pressure. The pale yellow oil which remains still contains a little thionyl chloride, and this is best removed by adding a further 100 ml. of anhydrous benzene and again distilling under reduced pressure. The residue is dissolved in 150 ml. of refluxing, anhydrous hexane (Note 2). The hot solution is treated with charcoal and filtered, and the filtrate is chilled to 0° in a sealed flask. The product, which crystallizes as colorless plates (Note 3), is filtered, washed with a little cold hexane, dried at 25° under vacuum for 2 hours, and stored in a tightly stoppered bottle. This gives 42–45 g., (77–84%) of product, m.p. 51–53°. Concentration of the hexane mother liquors to about 50 ml. followed by chilling to 0° and addition of a seed crystal gives a further 2.5–4.0 g. (5–8%) of product of equal purity. The total yield of diphenylacetyl chloride, m.p. 51–53°, is thus 44.5–49 g. (82–94%) (Note 4).

B. *Diphenylketene.* In a 500-ml. three-necked flask equipped with a magnetic stirring bar, a gas inlet tube, a calcium chloride drying tube, and a dropping funnel is placed a solution of 23.0 g. (0.10 mole) of diphenylacetyl chloride in 200 ml. of anhydrous ether. The flask is cooled in an ice-bath, dry nitrogen is passed through the system, and 10.1 g. (0.10 mole) of triethylamine is added dropwise during 30 minutes to the stirred solution. Triethylamine hydrochloride precipitates as a colorless solid, and the ether becomes bright yellow in color. When addition of the triethylamine is complete, the flask is tightly stoppered and stored overnight at 0°. The triethylamine hydrochloride is collected on a 9-cm. sintered glass funnel and washed with anhydrous ether until the washings are colorless. The ether is removed under reduced pressure and the residual red oil is transferred to a 50-ml. distilling flask fitted with a 10-cm. Vigreux column and distilled (Note 5). This gives 10.2–10.8 g. (53–57%) of diphenylketene as an orange oil, b.p. 118–120° (1 mm.) (Note 6). It can be stored at 0° in a tightly stoppered bottle for several weeks without decomposition.

2. Notes

1. Superior grade diphenylacetic acid, m.p. 147–148° (Matheson Coleman and Bell) was used without further purification.

2. Commercial hexane, A.C.S. grade (Matheson Coleman and Bell) was dried by distillation from potassium hydroxide.

3. Diphenylacetyl chloride crystallizes best when a seed crystal is added to the cold hexane solution. If, after several hours at 0°, crystallization has not commenced, scratching with a glass rod is sufficient to induce crystallization.

4. The checkers found it necessary to recrystallize the diphenylacetyl chloride twice to obtain the reported melting point.

5. Most of the diphenylketene distils cleanly at 118–119° (1 mm.) but strong heating is necessary for distillation of the final portion from the polymeric pot residue.

The checkers used a variety of different distillation set-ups.

The best results appeared to be obtained when the oil being distilled filled the flask to about two-thirds its capacity, the Vigreux column was no longer than 10 cm. in length, the whole apparatus was kept as small as possible, and the distillation was conducted as rapidly as possible.

6. The submitters obtained yields of 73–84% on the scale described and yields of up to 70% on a scale twice that described. From infrared analysis of the crude (undistilled) product the checkers concluded that this material represented a yield in excess of 80%. Thus the critical step appears to be the distillation. The checkers have used the crude (undistilled) product for some applications, but this procedure has not been uniformly successful and is not recommended.

3. Discussion

Diphenylacetyl chloride has been obtained from diphenylacetic acid with phosphorus pentachloride,[2] phosphorus oxychloride and phosphorus pentachloride,[3] and thionyl chloride.[4] It has also been prepared by treatment of diphenylketene with hydrogen chloride.[5] The methods of preparation of diphenylketene have been reviewed in an earlier procedure given in this series.[6] To those cited in the earlier procedure should be added the debromination of α-bromodiphenylacetyl bromide with triphenylphosphine.[7] The procedure above is a modification of that described by Staudinger.[8]

The present preparation consists of two very simple steps, uses relatively inexpensive starting materials, and does not involve hazardous or toxic chemicals or special apparatus. An important advantage is that the diphenylketene, until it is finally distilled, is never exposed to temperatures greater than 30–35°; hence polymerization is minimized (cf. ref. 6).

1. Department of Chemistry, Princeton University, Princeton, New Jersey.
2. F. Klingemann, *Justus Liebigs Ann. Chem.*, **275**, 50 (1893).
3. A. Bistrzycki and A. Landtwing, *Ber.*, **41**, 686 (1908).
4. W. A. Bonner and C. J. Collins, *J. Amer. Chem. Soc.*, **75**, 5372 (1953).
5. H. Staudinger, *Ber.*, **38**, 1735 (1905); *Justus Liebigs Ann. Chem.*, **356**, 51 (1907).
6. L. I. Smith and H. H. Hoehn, *Org. Syn.*, Coll. Vol. **3**, 356 (1955).
7. S. D. Darling and R. L. Kidwell, *J. Org. Chem.*, **33**, 3974 (1968).
8. H. Staudinger, *Ber.*, **44**, 1619 (1911).

THE FORMATION AND ALKYLATION OF SPECIFIC ENOLATE ANIONS FROM AN UNSYMMETRICAL KETONE: 2-BENZYL-2-METHYLCYCLOHEXANONE AND 2-BENZYL-6-METHYLCYCLOHEXANONE

(Cyclohexanone, 2-benzyl-2-methyl and cyclohexanone, 2-benzyl-6-methyl)

Submitted by Martin Gall and Herbert O. House[1]
Checked by K. E. Wilson and S. Masamune

1. Procedure

A. *1-Acetoxy-2-methylcyclohexene. Caution!* Since mixtures of perchloric acid with small amounts of organic material can explode violently, the perchloric acid should always be the last component added to the reaction mixture.

To a 1-l. flask are added 600 ml. of carbon tetrachloride, 250 ml. (2.7 moles) of acetic anhydride, 56 g. (0.50 mole) of 2-methylcyclohexanone, and 0.34 ml. (0.002 mole) of aqueous 70% perchloric acid. The reaction flask is stoppered and allowed to stand at room temperature for 3 hours during which time the reaction solution becomes first yellow-orange and finally red in color. The reaction mixture is poured into a cold (0–5°) mixture of 400 ml. of saturated aqueous sodium bicarbonate and 400 ml. of pentane which is contained in a 4-l. Erlenmeyer flask equipped with a mechanical stirrer. While the mixture is maintained at 0–5° and stirred vigorously, solid sodium bicarbonate is added in 3–5 g. portions as rapidly as foaming of the reaction mixture will permit. The addition of solid sodium bicarbonate is continued until all of the acetic acid has been neutralized and the aqueous phase remains slightly basic (pH 8). This neutralization requires approximately 400 g. of solid sodium bicarbonate which is added in portions over a period of approximately 3 hours. As soon as the neutralization is complete (Note 1), the organic layer (the lower layer) is separated and the aqueous phase is extracted with three 200-ml. portions of pentane. The combined organic solutions are dried over anhydrous magnesium sulfate and then concentrated by distilling the bulk of the pentane through a 30-cm. Vigreux column. The remaining solvents are removed by concentration under reduced pressure with a rotary evaporator and the residual liquid is distilled (Note 2) under reduced pressure. The 1-acetoxy-2-methylcyclohexene is collected as 66.6–70.9 g. (87–92%) of colorless liquid, b.p. 81–86° (18 mm.), $n^{25}D$ 1.4562–1.4572 (Note 3).

B. *2-Benzyl-2-methylcyclohexanone. Caution!* Ethereal solutions of methyllithium in contact with atmospheric oxygen may catch fire spontaneously. Therefore any manipulations with this reagent must be carried out with the utmost care to avoid accidental *spillage. Benzyl bromide is a powerful lachrymator. Steps B and C should be performed in an efficient fume hood.*

A 1-l. three-necked flask is equipped with a nitrogen-inlet tube fitted with a stopcock, a glass joint fitted with a rubber septum, a 125-ml. pressure-equalizing dropping funnel, a thermometer, and a glass-covered magnetic stirring bar. After the apparatus has been dried in an oven, 20 mg. of 2,2′-bipyridyl is added to the flask and the apparatus is thoroughly flushed with anhydrous, oxygen-free nitrogen (Note 4). A static nitrogen atmosphere is maintained in the reaction vessel throughout the subsequent operations involving organometallic reagents (Note 5). An ethereal solution containing 0.40 mole of methyllithium (Note 6) is added to the reaction vessel from a hypodermic syringe. The ether is removed by evacuating the apparatus while the solution is stirred and the flask is warmed with a water bath (40°) (Note 7). The reaction vessel is refilled with nitrogen and then 400 ml. of 1,2-dimethoxyethane (b.p. 83°, freshly distilled from lithium aluminum hydride) is transferred to the reaction vessel with a hypodermic syringe or a stainless steel cannula. The resulting purple solution of methyllithium and the methyllithium bipyridyl charge-transfer complex is cooled to 0–10° and then 29.3 g. (0.190 mole) of 1-acetoxy-2-methylcyclohexene is added, dropwise and with stirring, over a period of 35–45 minutes (Note 8) while the temperature of the reaction mixture is maintained at 0–10° with an ice bath. After the addition of the enol acetate, the reaction solution must still retain a light red-orange color indicating the presence of a small amount of excess methyllithium (Note 9). To this cold (10°) solution is added rapidly (10–15 seconds) with stirring, 68.4 g. (0.400 mole) of freshly distilled benzyl bromide [b.p. 78–79° (12 mm.), n^{25} D 1.5738]. The resulting yellow solution is stirred for 2–2.5 minutes (during which time the temperature of the reaction mixture rises from 10 to about 30°) and then poured into 500 ml. of cold (0–10°), saturated aqueous sodium bicarbonate and extracted with three 150-ml. portions of pentane. The combined organic extracts are dried over anhydrous magnesium sulfate and then concentrated under reduced pressure with a rotary evaporator. The residual liquid is fractionally distilled under reduced pressure to separate 31–41 g. of forerun fractions, b.p. 71–89° (20 mm.) and 41–87° (0.3 mm.) (Note 10), and 20.7–22.2 g. (54–58%) of 2-benzyl-2-methylcyclohexanone as a colorless to pale yellow liquid, b.p. 87–93° (0.3 mm.), n^{25} D 1.5322–1.5344 (Notes 11 and 12).

C. *2-Benzyl-6-methylcyclohexanone. Caution! The same precaution as that described in part B should be exercised in this step.*

A 1-l. three-necked flask is equipped as described in part B. After the assembled apparatus has been dried in an oven, 45 mg. of 2,2'-bipyridyl is added to the flask and the apparatus is thoroughly flushed with anhydrous, oxygen-free nitrogen (Note 4). A static nitrogen atmosphere is maintained in the reaction vessel throughout the subsequent operations involving organometallic reagents (Note 5). An ethereal solution containing 0.20 mole of methyllithium (Note 6) is added to the reaction flask with a hypodermic syringe. The ether is removed under reduced pressure as described in part B (Note 7). After the ether has been removed, the reaction vessel is refilled with nitrogen and then 400 ml. of 1,2-dimethoxyethane (b.p. 83°, freshly distilled from lithium aluminum hydride) is added to the reaction vessel with a hypodermic syringe or a stainless steel cannula. The resulting purple solution of methyllithium and the methyllithium-bipyridyl charge-transfer complex is cooled to −50° with a dry ice-methanol bath and then 21.0 g. (29 ml., 0.208 mole) of diisopropylamine (b.p. 84–85°, freshly distilled from calcium hydride) is added from a hypodermic syringe, dropwise and with stirring. During this addition, which requires 2–3 minutes, the temperature of the reaction solution should not be allowed to rise above −20° (Note 12). The resulting reddish-purple solution of lithium diisopropylamide and the bipyridyl charge-transfer complex is stirred at −20° for 2–3 minutes and then 50 ml. of a 1,2-dimethoxyethane solution containing 21.3 g. (0.190 mole) of 2-methylcyclohexanone is added, dropwise and with stirring. During this addition the temperature of the reaction solution should not be allowed to rise above 0° (Note 12). After the addition of the ketone, the resulting solution of the lithium enolate must still retain a pale reddish-purple color indicating the presence of a slight excess of lithium diisopropylamide (Notes 9 and 13). The solution of the lithium enolate is stirred and rapidly warmed to 30° with a water bath. Then 68.4 g. (0.400 mole) of freshly distilled benzyl bromide [b.p. 78–79° (12 mm.), n^{25} D 1.5738] is added from a hypodermic syringe, rapidly and with vigorous stirring. The resulting reaction causes the temperature of the reaction mixture to rise to about 50° within 2 minutes and then begin to fall. After a total reaction period of 6 minutes, the reaction mixture is poured into 500 ml. of cold (0–10°), saturated aqueous sodium bicarbonate and extracted with three 150-ml. portions of pentane. The combined organic extracts are washed successively with two 100-ml. portions of aqueous 5% hydrochloric acid and 100 ml. of saturated aqueous sodium bicarbonate and then dried over anhydrous magnesium sulfate and concentrated with a rotary evaporator. The residual yellow liquid is fractionally distilled under reduced pressure (Note 14) to separate 31–32 g. of forerun fractions, b.p. 67–92° (20 mm.) and 40–91° (0.3 mm.) (Note 10), and 21.3–23.3 g. (58–61%) of crude 2-benzyl-6-methylcyclohexanone as a colorless liquid, b.p. 91–97° (0.3 mm.), n^{25} D 1.5282–1.5360. The residue (10–11 g.) contains dibenzylated products. The crude reaction product contains (Notes 11 and 15) 2-benzyl-6-methylcyclohexanone (86–90%) and 2-benzyl-2-methylcyclohexanone (10–14%) accompanied in some cases by small amounts of *trans*-stilbene (Note 13).

To obtain the pure 2,6-isomer, the following procedure can be followed. After a 200-ml. flask, equipped with a Teflon-covered magnetic stirring bar, has been dried in an oven and flushed with nitrogen, 2.59 g. (0.0479 mole) of sodium methoxide (Note 16) is added, and the flask is stoppered with a rubber septum. A static nitrogen atmosphere is maintained in the reaction vessel throughout the remainder of the reaction. To the flask is added from a hypodermic syringe, 90 ml. of ether (freshly distilled from lithium aluminum hydride), and the resulting suspension is cooled with an ice bath. To the cold suspension is added from a hypodermic syringe a mixture of the crude distilled alkylated product (about 21–23 g.) and 3.74 g.

(0.0505 mole) of ethyl formate (Note 17). The mixture is stirred for 10 minutes with ice-bath cooling. The bath is then removed and stirring is continued for an additional 50 minutes. The resulting yellow suspension is treated with 300 ml. of water and then extracted with 250 ml. of ether. After the ethereal extract has been washed with 100 ml. of aqueous $1M$ sodium hydroxide, it is dried over anhydrous magnesium sulfate and concentrated with a rotary evaporator. The residual yellow liquid is distilled under reduced pressure to separate 16.2–17.3 g. (overall yield 42–45%) of pure (Note 11) 2-benzyl-6-methylcyclohexanone as a colorless liquid, b.p. 95–100° (0.3 mm.), n^{25} D 1.5299–1.5328.

2. Notes

1. Because the enol acetate is slowly hydrolyzed even by neutral aqueous solutions, the reaction mixture should be neutralized and the organic product should be separated and dried as rapidly as is practical.

2. The glassware employed in the distillation should be washed first with ammonium hydroxide and then water and finally dried in an oven before use to avoid the possibility of acid-catalyzed hydrolysis or rearrangement of the enol acetate during the distillation.

3. The submitters have been unsuccessful in finding a convenient gas chromatographic column which will separate 1-acetoxy-2-methylcyclohexene from its double-bond isomer, 1-acetoxy-6-methylcyclohexene. However, the n.m.r. spectrum (CCl_4 solution) of the product exhibits a peak at 2.02 p.p.m. (singlet, CH_3CO) superimposed on a multiplet in the region 1.3–2.2 p.p.m. (vinyl CH_3 and aliphatic CH_2) and lacks absorption at 0.98 p.p.m. where 1-acetoxy-6-methylcyclohexene exhibits a doublet ($J = 7$ Hz.) attributable to the aliphatic methyl group.[2] Consequently, the product contains less than 5% of the unwanted double bond isomer. The product exhibits infrared absorption (CCl_4 solution) at 1755 cm.$^{-1}$ (enol ester C=O) and 1705 cm.$^{-1}$ (C=C).

4. A good grade of commercial prepurified nitrogen can be used without further purification. A suitable method for the

purification of nitrogen is described by H. Metzer and E. Müller in "Methoden der Organischen Chemie," (Houben-Weyl) Vol. 1/2, 4th ed., Georg Thieme Verlag, Stuttgart, 1959, p. 327.

5. The apparatus illustrated in Figure 1 is convenient both for evacuating the reaction vessel and refilling it with nitrogen and also for maintaining a static atmosphere of nitrogen at slightly above atmospheric pressure in the reaction vessel.

6. A solution of methyllithium in ether was purchased from Alfa Inorganics, Inc. Directions for the preparation of methyllithium from methyl bromide are also available.[3] Solutions of methyllithium should be standardized immediately before use by the titration procedure of Watson and Eastham.[4] A standard $0.500M$ solution of 2-butanol (b.p. 99–100°, freshly distilled from calcium hydride) in p-xylene (b.p. 137–138°, freshly distilled from sodium) is prepared in a volumetric flask. A 25-ml. round-bottom flask, fitted with a rubber septum and a glass-covered magnetic stirring bar, is dried in an oven. After 1–2 mg. of 2,2'-bipyridyl has been added to the flask, it is flushed with anhydrous, oxygen-free nitrogen by inserting hypodermic needles through the rubber septum to allow gas to enter and escape. The tip of a 10-ml. burette is forced through the rubber septum and a measured volume of the standard 2-butanol

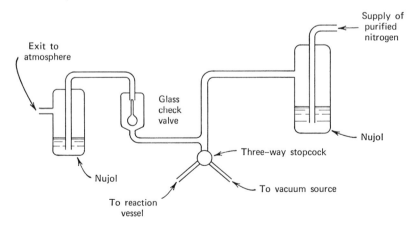

Figure 1. Apparatus for either evacuating or supplying a nitrogen atmosphere to the reaction vessel.

solution is added to the flask. Then 2.50 ml. of methyllithium solution is added to the reaction flask. The mixture is stirred and additional standard 2-butanol solution is added to the flask from the burette until the purple color of the methyllithium-bipyridyl complex is just discharged. For a 1.66M solution of methyllithium, 8.30 ml. of the standard 2-butanol solution is required in this titration.

7. Since a number of lithium enolates are significantly less soluble and less reactive in diethyl ether than in 1,2-dimethoxyethane, the submitters recommend the general use of this simple procedure to remove the diethyl ether before the lithium enolate is generated.

8. Lithium enolates react relatively slowly with enol acetates to form C-acetylated products. Consequently, the enol acetate should be added slowly with efficient stirring so that regions with high local concentrations of both the enolate anion and the enol acetate do not occur in the reaction solution.

9. It is important that the indicator color, showing a small excess of strong base, not be discharged completely since the presence of any excess enol acetate or ketone will permit equilibration of the isomeric metal enolates. Consequently, the addition of this reactant is complete if further additions will discharge completely the color of the indicator.

10. The various fractions of the forerun were analyzed employing a gas chromatography column packed with silicone gum, No. XE-60, suspended on Chromosorb P and heated to 248°. The components found (with the retention times indicated) were: benzyl bromide (9.0 minutes), 2-methylcyclohexanone (5.3 minutes), and, in some cases, bibenzyl (22.6 minutes). The bibenzyl, formed by reaction of the benzyl bromide with the excess methyllithium,[5] was identified from the infrared spectrum of a sample collected from the gas chromatograph.

11. On a 6-m. gas chromatography column packed with silicone gum, No. XE-60, suspended on Chromosorb P and heated to 240°, 2-benzyl-2-methylcyclohexanone (retention time 35.0 minutes) and 2-benzyl-6-methylcyclohexanone (retention time 33.2 minutes, *cis*- and *trans*-isomers not resolved) are partially resolved. However, the use of this analytical

method to detect small amounts of one structural isomer in the presence of the other is not reliable. Gas chromatography, however, can be used to determine the presence of any *trans*-stilbene (retention time 39.0 minutes) in the crude product.

The proportions of structurally isomeric benzylmethylcyclohexanones can be more accurately measured from the n.m.r. spectra of the distilled monoalkylated products. Pure 2-benzyl-6-methylcyclohexanone (principally the more stable *cis*-isomer in which both substituents are equatorial) exhibits the following n.m.r. absorptions (C_6D_6 solution): 7.0–7.3 (multiplet 5H, aryl CH), 2.9–3.5 (multiplet with at least 5 lines, 1H, one of the two nonequivalent benzylic CH_2 protons), 1.1–2.6 (multiplet, 9H, aliphatic CH and the second of the two nonequivalent benzylic CH_2 protons), and 0.97 p.p.m. (doublet, $J = 6.0$ Hz., 3H, CH_3). In carbon tetrachloride solution the corresponding peaks are found at 7.0–7.3, 2.9–3.5, 1.1–2.7, and 0.97 p.p.m. (doublet, $J = 6.0$ Hz.). In this solvent a second weak doublet ($J = 6.5$ Hz.) is present at 1.04 p.p.m. and is attributable to the small amount of the less stable *trans*-2-benzyl-6-methylcyclohexanone (one equatorial and one axial substituent) present. The 2,6-isomer exhibits infrared absorption (CCl_4 solution) at 1710 cm.$^{-1}$ (C=O) and a series of weak (ϵ 202 to 335) ultraviolet maxima (95% EtOH solution) in the region 240–270 mμ. The mass spectrum exhibits a molecular ion at m/e 202 with relatively abundant fragment peaks at m/e 159, 145, 117, 111, and 91 (base peak). Pure 2-benzyl-2-methylcyclohexanone has the following n.m.r. absorptions (C_6D_6 solution): 6.9–7.3 (multiplet, 5H, aryl CH), 2.78 (singlet, 2H, benzylic CH_2), 2.1–2.4 (multiplet, 2H, CH_2CO), 1.2–1.7 (multiplet, 6H, aliphatic CH), and 0.91 p.p.m. (singlet, 3H, CH_3). In carbon tetrachloride solution the corresponding peaks are found at 6.9–7.3, 2.78, 2.2–2.6, 1.4–2.0, and 0.95 p.p.m. This ketone has infrared absorption (CCl_4 solution) at 1710 cm.$^{-1}$ (C=O) and shows a series of weak (ϵ 140 to 284) ultraviolet maxima (95% EtOH solution) in the region 240–270 mμ. The mass spectrum exhibits a molecular ion at m/e 202 with relatively abundant fragment peaks at m/e 159, 117, 92, 91 (base peak), 55, 44, 43, and 41. Mixtures of the 2,6- and 2,2-isomers could be analyzed by measuring their n.m.r. spectra in deuteriobenzene solution and integrating the region 2.6–3.5 p.p.m. The peak at 2.78 p.p.m., attributable to both benzylic hydrogen atoms of the 2,2-isomer, is well resolved from the multiplet at 2.9–3.5 p.p.m., attributable to one of the two benzylic hydrogen atoms of the 2,6-isomer. Utilizing this method (which is in agreement with the less reliable value obtained by gas chromatographic analysis), no 2,6-isomer is detected in the 2-benzyl-2-methylcyclohexanone prepared by the present procedure. The 2-benzyl-6-methylcyclohexanone product contains 10–14% of the 2,2-isomer.

12. At temperatures above 0°, 1,2-dimethoxyethane is slowly attacked by lithium diisopropylamide resulting in the protonation of the strong base.

13. If this precaution is not followed, partial or complete equilibration of the enolates will occur because of proton transfers between the enolates and the excess un-ionized ketone. In an experiment where a slight excess of ketone was added, the distilled, monoalkylated product (40% yield) contained 77% of the undesired 2,2-isomer and only 23% of the desired 2,6-isomer. However, it is also important in this preparation not to allow a large excess of lithium diisopropylamide to remain in the reaction mixture; this base reacts with benzyl bromide to form *trans*-stilbene[6] which is difficult to separate from the reaction product.

14. During the early part of the distillation when a substantial amount of benzyl bromide is present, the still pot should not be heated above 140° to avoid serious discoloration. When the bulk of the benzyl bromide has been removed, the temperature of the still pot may be raised to 150–160° to facilitate distillation of the product.

15. The proportions of the desired 2,6-isomer and the unwanted 2,2-isomer in the alkylated product will vary depending on the rate and efficiency of mixing of the benzyl bromide with the lithium enolate. If the alkylation of the initially formed enolate could be effected without any enolate equilibration, less than 2% of the unwanted 2,2-isomer would be expected.[7]

16. The sodium methoxide was purchased from Matheson

Coleman and Bell. Material from a freshly opened bottle was used without further purification.

17. Commercial ethyl formate (Eastman Organic Chemicals) was purified by stirring it successively over anhydrous sodium carbonate and over anhydrous magnesium sulfate. The material was distilled to separate pure ethyl formate, b.p. 54–54.5°.

3. Discussion

2-Benzyl-6-methylcyclohexanone has been prepared by the hydrogenation of 2-benzylidene-6-methylcyclohexanone over a platinum or nickel catalyst,[8] and by the alkylation of the sodium enolate of 2-formyl-6-methylcyclohexanone with benzyl iodide followed by cleavage of the formyl group with aqueous base.[9] The 2,6-isomer was also obtained as a minor product (about 10% of the monoalkylated product) along with the major product, 2-benzyl-2-methylcyclohexanone by successive treatment of 2-methylcyclohexanone with sodium amide and then with benzyl chloride or benzyl bromide.[10,11] Reaction of the sodium enolate of 2-formyl-6-methylcyclohexanone with potassium amide in liquid ammonia formed the corresponding dianion which was first treated with 1 equiv. of benzyl chloride and then deformylated with aqueous base to form 2-benzyl-2-methylcyclohexanone.[12]

These synthetic routes illustrate well the classical methods which have been used for the alkylation of unsymmetrical ketones. Reaction of the ketone with a strong base such as sodium amide under conditions which permit equilibration of the enolates affords an equilibrium mixture of enolates and subsequent reaction with an alkylating agent yields a mixture of monoalkylated products as well as polyalkylated products. In the present case, the equilibrium mixture of metal enolates from 2-methylcyclohexanone contains 10–35% of the less highly substituted double bond isomer.[2,13,14] Consequently the major alkylation product from this mixture is the 2,2-isomer. If the methylene group is protected with a blocking group, the resulting ketone is alkylated solely at the more highly substituted alpha carbon. Removal of the blocking group affords pure 2,2-isomer. Alternatively, an activating group such as a formyl group or a carboalkoxyl group can be introduced at the less highly substituted alpha carbon to permit selective alkylation at this position; the activating group is then removed.

An alternative solution to the problem of effecting the selective alkylation of an unsymmetrical ketone consists of generating a specific enolate under conditions where the enolate isomers do not equilibrate.[15] The methods which have been used to generate specific enolate anions include the reduction of enones[16,17] or α-haloketones[18,19] with metals, the reaction of organolithium reagents with enol silyl ethers[7,20,21] or enol esters,[15,21] and the kinetically controlled abstraction of the least hindered alpha proton from a ketone with a hindered base such as lithium diisopropylamide.[7] The present procedures illustrate the last two methods. To prevent the equilibration of lithium enolates during their formation, care is taken that no proton-donor BH (such as an alcohol or the un-ionized ketone) is present. Although with attention to this precaution, either of the structurally isomeric enolate ions can be prepared and maintained in solution, this fact does not ensure a structurally specific alkylation. As the accompanying equations illustrate, once reaction of the enolate with the alkyl halide is initiated, the reaction mixture will necessarily contain an un-ionized ketone, namely the alkylated product, and equilibration of the enolate ions can occur. Consequently, a structurally specific alkylation of an enolate anion can be successful only if the alkylation reaction is more rapid than equilibration so that the starting enolate **2** is consumed by the alkylating agent before significant amounts of the unwanted enolate **1** have been formed. In practice, this criterion normally is fulfilled with very reactive alkylating agents such as methyl iodide. With less reactive alkylating agents such as benzyl bromide and n-alkyl iodides, some equilibration is usually observed.[21] The problem is aggravated when the alkylation involves the less stable and/or the less reactive enolate isomer (*e.g.*, **2**). In the present procedures, relatively high concentrations of the enolate and the alkyl halide are employed to increase the alkylation rate and, consequently, to decrease the proportion of the unwanted monoalkylation product which results from equilibration prior to alkylation. As is to be expected from the foregoing discussion, the alkylation of the enolate **1** to form 2-benzyl-2-methylcyclohexanone exhibits more structural specificity than the alkylation of the enolate **2** to form 2-benzyl-6-methylcyclohexanone. The unwanted 2,2-isomer (10–14%) which accompanies 2-benzyl-6-methylcyclohexanone in this alkylation is removed by a well-known chemical separation procedure in which the 2,2-isomer is converted to its formyl derivative.[22,23]

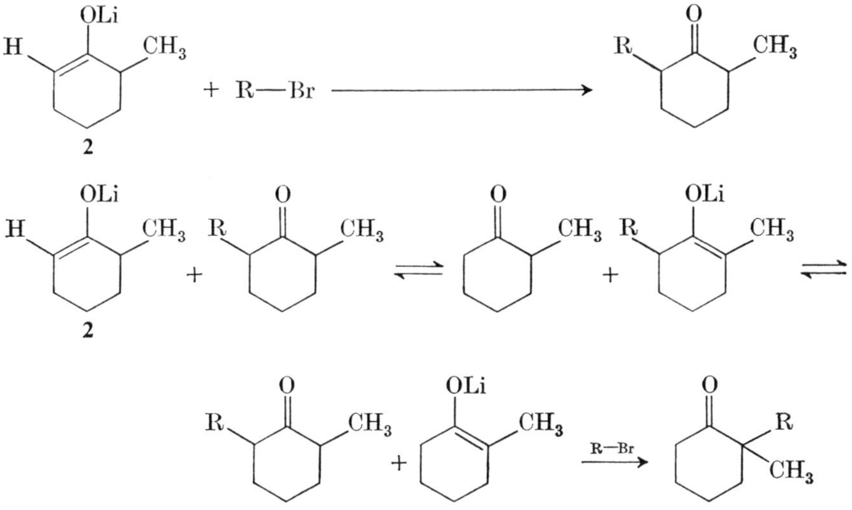

1. School of Chemistry, Georgia Institute of Technology, Atlanta, Georgia 30332.
2. H. O. House and V. Kramar, *J. Org. Chem.*, **28**, 3362 (1963).
3. G. Wittig and A. Hesse, *Org. Syn.*, **50**, 66 (1970).
4. S. C. Watson and J. F. Eastham, *J. Organometal. Chem.*, **9**, 165 (1967).
5. H. Gilman and F. K. Cartledge, *J. Organometal. Chem.*, **2**, 447 (1964); H. Gilman and A. H. Haubein, *J. Amer. Chem. Soc.*, **66**, 1515 (1944).

6. C. R. Hauser, W. R. Brasen, P. S. Skell, S. W. Kantor, and A. E. Brodhag, *J. Amer. Chem. Soc.*, **78**, 1653 (1956).
7. H. O. House, L. J. Czuba, M. Gall, and H. D. Olmstead, *J. Org. Chem.*, **34**, 2324 (1969).
8. R. Cornubert and C. Borrel, *Compt. Rend.*, **183**, 294 (1926); *Bull. Soc. Chim. Fr.*, **46**, 1148 (1929); P. Anziani, A. Aubrey, P. Bourguignon, and R. Cornubert, *Bull. Soc. Chim. Fr.*, 1202 (1950).
9. H. K. Sen and K. Mondal, *J. Indian Chem. Soc.*, **5**, 609 (1928).
10. R. Cornubert and H. Le Bihan, *Compt. Rend.*, **186**, 1126 (1928).
11. R. Cornubert, C. Borrel, and H. Le Bihan, *Bull. Soc. Chim. Fr.*, **49**, 1381 (1931).
12. S. Boatman, T. M. Harris, and C. R. Hauser, *J. Amer. Chem. Soc.*, **87**, 82 (1965).
13. H. O. House, W. L. Roelofs, and B. M. Trost, *J. Org. Chem.*, **31**, 646 (1966).
14. D. Caine, *J. Org. Chem.*, **29**, 1868 (1964); D. Caine and B. J. L. Huff, *Tetrahedron Lett.*, 4695 (1966).
15. For a brief review, see H. O. House, *Rec. Chem. Progr.*, **28**, 99 (1967).
16. G. Stork, P. Rosen, N. Goldman, R. V. Coombs, and J. Tsuji, *J. Amer. Chem. Soc.*, **87**, 275 (1965).
17. H. A. Smith, B. J. L. Huff, W. J. Powers, III, and D. Caine, *J. Org. Chem.*, **32**, 2851 (1967); L. E. Hightower, L. R. Glasgow, K. M. Stone, D. A. Albertson, and H. A. Smith, *J. Org. Chem.*, **35**, 1881 (1970).
18. M. J. Weiss, R. E. Schaub, G. R. Allen, J. F. Poletto, C. Pidacks, R. B. Conrow, and.C. J. Coscia, *Tetrahedron*, **20**, 357 (1964).
19. T. A. Spencer, R. W. Britton, and D. S. Watt, *J. Amer. Chem. Soc.*, **89**, 5727 (1967).
20. G. Stork and P. F. Hudrlik, *J. Amer. Chem. Soc.*, **90**, 4462, 4464 (1968).
21. H. O. House and B. M. Trost, *J. Org. Chem.*, **30**, 2402 (1965); H. O. House, M. Gall, and H. D. Olmstead, *J. Org. Chem.*, **36**, 2361 (1971).
22. W. J. Bailey and M. Madoff, *J. Amer. Chem. Soc.*, **76**, 2707 (1954); F. E. King, T. J. King, and J. G. Topliss, *J. Chem. Soc.*, 919 (1957).
23. The application of this procedure to the separation of the benzylmethylcyclohexanone isomers was developed by Michael J. Umen.

FORMATION AND PHOTOCHEMICAL WOLFF REARRANGEMENT OF CYCLIC α-DIAZO KETONES: D-NORANDROST-5-EN-3β-OL-16-CARBOXYLIC ACIDS

(D-Norandrost-5-ene-16-carboxylic acids, 3β-hydroxy-)

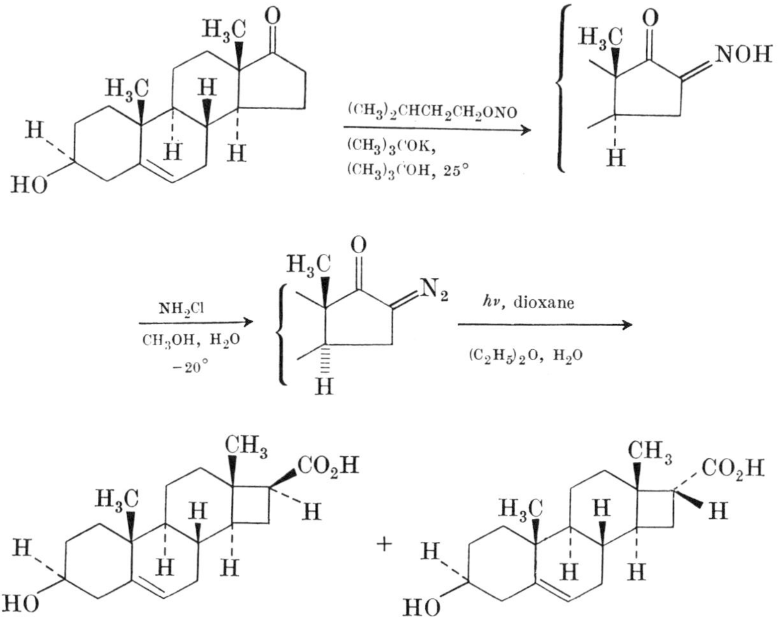

Submitted by Thomas N. Wheeler and J. Meinwald[1]
Checked by R. A. Blattel, D. G. B. Boocock, and Peter Yates

1. Procedure

A. *16-Oximinoandrost-5-en-3β-ol-17-one.* A 2-l. three-necked, round-bottomed flask is fitted with a reflux condenser, a mechanical stirrer, and a pressure-equalizing dropping funnel. To the reaction flask is added 750 ml. of anhydrous *t*-butyl alcohol (Note 1). As the *t*-butyl alcohol is slowly stirred, a stream of dry nitrogen is passed through the flask and 12.2 g. (0.312 mole) of potassium metal is added cautiously. The flask is surrounded by a water bath maintained at 70° to assist in dissolving the potassium metal. After 1.5 hours the stirred mixture is homogeneous. The water bath is removed, and the reaction mixture is allowed to cool to room temperature. To the potassium t-butoxide solution is slowly added 45.0 g. (0.156 mole) of dehydroisoandrosterone (Note 2), and stirring is continued for one hour until the gold-colored mixture is again homogeneous. To the reaction mixture is now added, dropwise, 42.0 ml. (36.5 g, 0.312 mole) of isoamyl nitrite (Note 3), and stirring is continued overnight at room temperature.

The deep orange reaction mixture is diluted with an equal volume of water, poured into a 2-l. separatory funnel, and acidified with aqueous $3M$ hydrochloric acid. The addition of 400 ml. of ether assists in effecting the separation of the clear yellow, aqueous, lower layer from the fluffy-white ethereal suspension that forms the upper layer. This suspension is filtered through a 250-ml. coarse sintered glass funnel, and the precipitate of oximino ketone is washed with ether several times. After drying overnight in a vacuum desiccator at $-5°$, 48.0–48.5 g. of a white product, m.p. 245–247° dec., is obtained, whose n.m.r. spectrum (pyridine solution) shows it to be a 1:1 solvate of the oximino ketone with t-butyl alcohol (Note 4); the yield is 79% (Notes 5 and 6). This product is used without further purification in the synthesis of the α-diazo ketone (Note 7).

B. *16-Diazoandrost-5-en-3β-ol-17-one.* A 1-l. three-necked, round-bottomed flask is fitted with a mechanical stirrer, a 50-ml. pressure-equalizing dropping funnel, and a thermometer. As stirring is initiated, 375 ml. of methanol and 72 ml. of aqueous $5M$ sodium hydroxide (0.36 mole) is added to the flask, followed by 18.0 g. (0.0460 mole) of the 1:1 solvate of 16-oximinoandrost-5-en-3β-ol-17-one with t-butyl alcohol. The oximino ketone readily dissolves to give a yellow solution. To the reaction mixture is added 28.3 ml. of concentrated aqueous ammonia (0.425 mole), and the flask is surrounded by an ice bath to maintain the reaction temperature at 20°. Through the dropping funnel 133 ml. of aqueous $3.0M$ sodium hypochlorite (0.40 mole) is added dropwise. The sodium hypochlorite solution should be kept near 0°, so 25-ml. portions should be added to the addition funnel and the remaining solution should be kept in an ice bath (Note 8). It is important that the rate of addition of the sodium hypochlorite and the position of the ice bath be adjusted so as to maintain the temperature of the reaction mixture at 20° ± 1° (Note 9). As soon as all of the sodium hypochlorite has been added, the ice bath is removed, and the reaction mixture is allowed to warm to room temperature and stirred for 6 hours.

The reaction mixture is diluted with an equal volume of water and extracted with a 400-ml. and a 200-ml. portion of methylene chloride. The combined methylene chloride extracts are washed with three 250-ml. portions of aqueous 20% sodium chloride, dried over anhydrous magnesium sulfate, and concentrated to leave a yellow solid. Recrystallization from acetone gives 8.0–9.3 g. (55–64%) of crystalline α-diazo ketone, m.p. 200–202° dec. (Note 10).

C. D-*Norandrost-5-en-3β-ol-16α- and 16β-carboxylic acids*. In a solution of 500 ml. of 1,4-dioxane, 1250 ml. of ether, and 250 ml. of water contained in a 3-l. three-necked, round-bottomed flask is dissolved 7.50 g. (0.0239 mole) of 16-diazoandrost-5-en-3β-ol-17-one. The flask is fitted with a reflux condenser, a quartz immersion well, and a nitrogen inlet. After the reaction vessel has been flushed with nitrogen, the diazo ketone solution is irradiated for 48 hours with a 450-watt Hanovia lamp with a Corex filter (Note 11). The photolysis mixture is decanted in portions into a 2-l. separatory funnel, washed three times with 500-ml. portions of water to remove the dioxane, and dried over magnesium sulfate. The ether is evaporated to leave a pale yellow residue. The residue is digested with 125 ml. of boiling methylene chloride under reflux for 30 minutes. The methylene chloride solution is allowed to cool to room temperature and filtered to separate about 1.4 g. of the crude α-isomer as a white powder. This solid is recrystallized by dissolving it in a large volume of methanol (125 ml.) and concentrating the solution to a small volume (25 ml.) to yield 1.2 g. (17%) of D-norandrost-5-en-3β-ol-16α-carboxylic acid as a white solid, m.p. 271–274° (Note 12). The β-isomer is most readily obtained by concentrating the methylene chloride mother liquor and dissolving the residue in a mixture of 75 ml. of methanol and 25 ml. of ether. This solution is treated with an excess of diazomethane in ether at room temperature. After one hour at room temperature, the excess diazomethane is removed with a stream of nitrogen, the solvent is evaporated, and the solid residue is chromatographed on 175 g. of Woelm neutral alumina Activity Grade II. Elution with a benzene-ether mixture (3:1 v/v) gives 3.9 g. of a white solid. This solid is recrystallized from ether-heptane to give 3.0–3.1 g. (39–41%) of white, crystalline methyl D-norandrost-5-en-3β-ol-16β-carboxylate, m.p. 161–163° (Notes 13 and 14).

2. Notes

1. Anhydrous t-butyl alcohol may be conveniently prepared by distilling it from calcium hydride into a receiver containing Type 4A molecular sieves.

2. Available from Searle Chemicals, Inc.

3. Isoamyl nitrite of sufficient purity may be prepared by the method of Noyes.[2] The isoamyl nitrite is stored over anhydrous magnesium sulfate until used.

4. The n.m.r. spectrum (pyridine solution) included signals at 0.97 (singlet, 3H), 1.03 (singlet, 3H), 1.42 (singlet, 9H) and 5.37 p.p.m. (multiplet, 1H).

5. The submitters, working on twice the scale described, obtained 90.0 g. (74%) of the solvate, m.p. 245–247°.

6. This oximino ketone has been previously prepared by a somewhat different procedure[3] and recrystallized from isopropyl alcohol to give crystals, m.p. 247–248°.

7. Other oximino ketones may be too soluble in ether to permit utilization of this isolation procedure. In this case, the submitters, working on twice the scale described, utilized the following procedure. After acidification of the reaction mixture, the oximino ketone is extracted into 1000 ml. of ether, and the ethereal extract is washed with four 100-ml. portions of saturated aqueous sodium bicarbonate and exhaustively extracted with aqueous 0.5M potassium hydroxide in 250-ml. portions until acidification of the basic extract gives no oximino ketone. A stream of nitrogen is bubbled through

the combined basic extracts to remove any dissolved ether, and the solution is then cooled in an ice bath and acidified with aqueous 3M hydrochloric acid. The precipitate is collected by suction filtration and dried in a vacuum desiccator.

8. A procedure for preparing concentrated sodium hypochlorite solution is given by Coleman and Johnson.[4] Common bleach solution, such as Clorox, may also be used, although the volume of the solution is considerably increased.

9. If the temperature of the reaction mixture is maintained below 20°, an appreciable amount of colorless α-dichloro ketone is obtained.[5] If the temperature is allowed to rise above 20°, the chloramine decomposes before it has time to react with the oximino ketone. The generation of chloramine *in situ* is quite exothermic, and care must be taken to maintain the temperature at 20°.

10. The submitters, working on twice the scale described, obtained 21.6 g. (74%) of product, m.p. 201–202° (dec.).

11. A Pyrex filter has been used; however, the photolysis appears to proceed more cleanly through Corex. The photolysis is considered complete when the infrared spectrum of a sample shows no diazo absorption at 2065 cm^{-1}.

12. Utilizing a somewhat different procedure, Mateos and Pozas have also obtained the α-carboxylic acid, m.p. 272–275°.[6]

13. The reported melting point of the β-methyl ester is 163–164°.[6]

14. The submitters, working on twice the scale described, obtained 2.8 g. (19%) of the α-acid, m.p. 272–275°, and 9.2 g (61%) of the methyl ester of the β-acid, m.p. 161–163°.

3. Discussion

The earliest methods for preparing cyclic α-diazo ketones involved the oxidation of the monohydrazones prepared from α-diketones, generally using mercuric oxide.[7,8] Recent modifications of this procedure include the use of calcium hypochlorite in aqueous sodium hydroxide or "activated" manganese dioxide as oxidants.[9] The latter reagent, especially, seems preferable to mercuric oxide. The base-catalyzed decomposition of the monotosylhydrazones of α-diketones has been used to prepare α-diazo ketones. Such reactions have been performed in aqueous sodium hydroxide,[10,11] with basic aluminum oxide in methylene chloride,[12] and with a variety of other bases. A promising and novel approach to cyclic α-diazo ketones involves the reaction of α-hydroxymethylene ketones with diethylamine and tosyl azide to give high yields of the α-diazo ketone.[13]

The present procedure for the synthesis of an α-diazo ketone is a modification of the Forster reaction,[14] which has been recently exploited by numerous workers.[11,15-19] The synthesis is generally applicable to cyclic ketones, is convenient, and offers moderate yields (60–70%) of pure α-diazo ketones.

The photochemical Wolff rearrangement represents a generally useful ring contraction technique.[20,21]

1. Department of Chemistry, Cornell University, Ithaca, New York 14850.
2. W. A. Noyes, *Org. Syn.*, Coll. Vol. **2**, 108 (1943).
3. F. Stodola, E. C. Kendall, and B. F. McKenzie, *J. Org. Chem.*, **6**, 841 (1941).
4. G. H. Coleman and H. L. Johnson, *Inorg. Syn.*, **1**, 59 (1939).
5. T. N. Wheeler, Ph.D. Thesis, Cornell University, 1969.
6. J. L. Mateos and R. Pozas, *Steroids*, **2**, 527 (1963).
7. T. Curtius and K. Thun, *J. Prakt. Chem.*, (2)**44**, 161 (1891).
8. C. D. Nenitzescu and E. Solomonica, *Org. Syn.*, Coll. Vol. **2**, 496 (1943).
9. H. Morrison, S. Danishefsky, and P. Yates, *J. Org. Chem.*, **26**, 2617 (1961).
10. M. P. Cava and R. L. Little, *Chem. and Ind. (London)*, 367 (1957).
11. M. P. Cava, R. L. Little, and D. R. Napier, *J. Amer. Chem. Soc.*, **80**, 2257 (1958).
12. J. M. Muchowski, *Tetrahedron Lett.*, 1773 (1966); J. K. Crandall, Ph.D. Thesis, Cornell University, 1963.
13. M. Rosenberger, P. Yates, J. B. Hendrickson, and W. Wolf, *Tetrahedron Lett.*, 2285 (1964); M. Regitz, *Angew Chem. Int. Ed. Engl.*, **6**, 733 (1967); J. B. Hendrickson and W. A. Wolf, *J. Org. Chem.*, **33**, 3610 (1968).
14. M. O. Forster, *J. Chem. Soc.*, **107**, 260 (1915).
15. J. Meinwald, G. G. Curtis, and P. G. Gassman, *J. Amer. Chem. Soc.*, **84**, 116 (1962).
16. M. P. Cava and E. Moroz, *J. Amer. Chem. Soc.*, **84**, 115 (1962).
17. G. Muller, C. Huynh, and J. Mathieu, *Bull. Soc. Chim. Fr.*, 296 (1962).
18. J. L. Mateos and O. Chao, *Bol. Inst. Quim. Univ. Nacl. Auton. Mex.*, **13**, 3 (1961) [*C.A.*, **57**, 9914 (1962)].
19. "The Synthesis, Properties, and Reactions of Several Steroidal α-Oximino and α-Diazoketones," University Microfilms, Ann Arbor, Mich., Order No. 65-3941; *Dissertation Abstr.*, (11) **25**, 6246 (1965).
20. F. Weygand and H. J. Bestmann, *Angew. Chem.*, **72**, 535 (1960).
21. W. Reid and H. Mengler, *Fortschr. Chem. Forsch.*, **5**, 1 (1965).

HYDROBORATION OF OLEFINS: (+)-ISOPINOCAMPHEOL
[(+)-3-Pinanol]

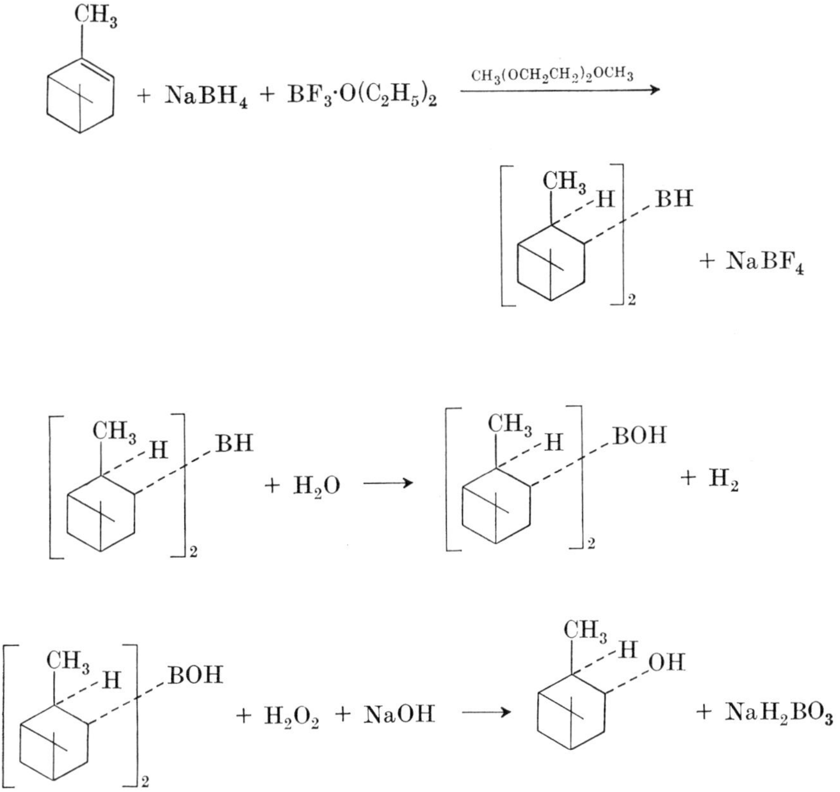

$$\text{α-pinene} + \text{NaBH}_4 + \text{BF}_3\cdot\text{O}(\text{C}_2\text{H}_5)_2 \xrightarrow{\text{CH}_3(\text{OCH}_2\text{CH}_2)_2\text{OCH}_3} [\text{pinanyl}]_2\text{BH} + \text{NaBF}_4$$

$$[\text{pinanyl}]_2\text{BH} + \text{H}_2\text{O} \longrightarrow [\text{pinanyl}]_2\text{BOH} + \text{H}_2$$

$$[\text{pinanyl}]_2\text{BOH} + \text{H}_2\text{O}_2 + \text{NaOH} \longrightarrow \text{isopinocampheol} + \text{NaH}_2\text{BO}_3$$

Submitted by G. Zweifel[1] and H. C. Brown
Checked by E. J. Corey, D. Shore, and Ravi K. Varma

1. Procedure

In a 300-ml. three-necked flask equipped with a condenser fitted with a calcium chloride tube, a pressure-equalizing dropping funnel, a thermometer, and a mechanical stirrer

(Note 1) are placed 3.1 g. (0.080 mole) of sodium borohydride, 100 ml. of diglyme (Note 2), and 27.2 g. (0.200 mole) of (−)-α-pinene (Note 3) diluted with 20 ml. of diglyme. The flask is immersed in a water bath (20–25°). Diborane is generated by dropwise addition of 14 ml. (0.11 mole, D^{20} 1.125) of boron trifluoride etherate (Note 4) to the well-stirred reaction mixture over a period of 15 minutes. The diisopinocampheylborane precipitates as white solid as the reaction proceeds. The mixture is maintained for an additional hour at room temperature. The excess hydride is then decomposed by dropwise addition of 20 ml. of water (Note 5). The organoborane formed (R_2BOH) is oxidized at 30–50° (water bath) by adding in one portion 22 ml. of aqueous $3M$ sodium hydroxide, followed by the dropwise addition of 22 ml. of aqueous 30% hydrogen peroxide to the well stirred reaction mixture. The flask is kept for an additional 30 minutes at room temperature.

The alcohol reaction mixture is extracted with 200 ml. of ether and the ether extract is washed five times with equal volumes of ice water to remove the diglyme. The ether extract is dried over anhydrous magnesium sulfate and the ether is removed by distillation through a short Vigreux column. The residue is distilled under reduced pressure to separate 26.2 g. (85%) of isopinocampheol, b.p. 80–82° (2 mm.). The distillate crystallizes immediately in the collection flask (Note 6). The crystals melt at 50–52° (Note 7). Recrystallization from about 10 ml. of petroleum ether (b.p. 35–37°) affords pure isopinocampheol as needles, m.p. 55–57°, $[\alpha]^{20}_D$ +32.8° (c, 10 in benzene).

2. Notes

1. The apparatus is dried in an oven and assembled under dry nitrogen. Alternatively it can be flamed dry in a stream of dry nitrogen. A magnetic stirrer can be utilized for small scale experiments.

2. Diglyme (dimethyl ether of diethylene glycol, b.p. 162° at 760 mm.) from Ansul Chemical Company, Marinette, Wisconsin, is purified in the following way. Diglyme (1 l.) is stored over 10 g. of granular calcium hydride for 12 hours. The diglyme is

decanted into a distilling flask and sufficient lithium aluminum hydride is added to ensure an excess of active hydride. The solvent is distilled at 62–63° (15 mm.).

3. (−)-α-Pinene ($[\alpha]^{20}$ D −47.9°) was obtained by the submitters from the Glidden Co., Jacksonville, Florida. Alternatively (−)-β-pinene ($[\alpha]^{20}$ D −21.1°) is readily available and can be isomerized to (−)-α-pinene by shaking with a palladium hydrogenation catalyst in the presence of hydrogen.[2] The checkers used (−)-α-pinene ($[\alpha]^{28}$ D −52.2°) supplied by Chemical Samples Co., 4692 Kenny Road, Columbus, Ohio, 43220.

4. Commercial boron trifluoride ethyl etherate (from Matheson Coleman and Bell) is distilled in an all glass apparatus at 46° (10 mm.) from a few pieces of granular calcium hydride. The latter serves to remove small quantities of volatile acids and greatly reduces bumping during distillation.

5. Since the olefin is hydroborated to the dialkylborane stage (R_2BH), a large amount of hydrogen is evolved on hydrolysis. Consequently, the addition of water should be carried out slowly (dropwise) and adequate ventilation is recommended.

6. For the distillation of the isopinocampheol the Vigreux column is attached to an air condenser. The receiver is immersed in an ice bath.

7. The checkers obtained from (−)-α-pinene, after distillation, (+)-isopinocampheol, m.p. 51–53° (corr.), $[\alpha]^{28}$ D +32.1°.

3. Discussion

(−)-Isopinocampheol has been prepared by hydrogenation of *trans*-pinocarveol with a neutral nickel catalyst at 70–100°.[3]

The hydroboration reaction provides a convenient procedure for the conversion of olefins to alcohols without rearrangement and with a predictable stereochemistry. The reaction has been applied to a large number of olefins of widely different structures.[4,5] The results obtained support the proposed generalization that hydroboration involves an *anti*-Markownikoff, *cis*-addition of borane from the less hindered side of the double bond.[6]

1. Department of Chemistry, University of California, Davis, California.
2. I. Richter and W. Wolff, *Ber.*, **59**, 1733 (1926).
3. H. Schmidt, *Ber.*, **77**, 544 (1944).
4. H. C. Brown, "Hydroboration," W. A. Benjamin, Inc., New York, 1962.
5. G. Zweifel and H. C. Brown, *Org. React.*, **13**, 1 (1963).
6. H. C. Brown and G. Zweifel, *J. Amer. Chem. Soc.*, **83**, 2544 (1961).

HYDROGENOLYSIS OF CARBON-HALOGEN BONDS WITH CHROMIUM(II)-EN PERCHLORATE: NAPHTHALENE FROM 1-BROMONAPHTHALENE

$$\text{1-BrC}_{10}\text{H}_7 + 2\,\text{Cr}^{II}(\text{en})_x + \text{H}_2\text{O} \xrightarrow[(\text{CH}_3)_2\text{NCHO}]{25°} \text{C}_{10}\text{H}_8 + 2\,\text{Cr}^{III}(\text{en})_x + \text{Br}^- + \text{HO}^-$$

Submitted by RUTH S. WADE and C. E. CASTRO[1]
Checked by NORTON P. PEET and HERBERT O. HOUSE

1. Procedure

A 250-ml. three-necked flask, equipped with a magnetic stirring bar and nitrogen inlet and outlet stopcocks, is charged with 60 ml. of dimethylformamide (Note 1) and 6.01 g. (0.100 mole) of ethylenediamine (Note 2). The outlet stopcock is connected to a trap containing mercury or Nujol, and the third neck of the flask is fitted with a rubber septum. While the solution in the reaction flask is stirred, the system is flushed with nitrogen for 30 minutes, and then a static nitrogen atmosphere is maintained in the reaction vessel during the remainder of the reaction. An aqueous solution containing 0.03 mole of chromium(II) perchlorate (Note 3) is added to the reaction vessel with a hypodermic syringe to form a purple solution of the chromium(II)-en complex. To this solution is added, with a hypodermic syringe, a solution of 1.66 g. (0.0080 mole) of 1-bromonaphthalene (Note 4) in 20 ml. of oxygen-free (Note 5) dimethylformamide (Note 1). The reaction solution is stirred for 70 minutes (Note 6), during which time the color changes from purple to deep red, and then poured into a solution of 40 g. of ammonium sulfate in 400 ml. of aqueous $0.4M$ hydrochloric acid. The resulting emulsion is extracted with five 60-ml. portions of ether, and the combined ethereal extracts are washed with two 25-ml. portions of water and then dried over potassium carbonate and concentrated. The residue crystallizes as 0.96–1.00 g. (93–98%) of naphthalene, m.p. 77–80°. Recrystallization from ethanol affords the pure naphthalene as white plates, m.p. 80–81°.

2. Notes

1. Baker reagent grade dimethylformamide was used without purification.

2. The submitters employed, without purification, 98% ethylenediamine obtained from Mallinckrodt Chemical Works; the checkers employed material from Eastman Organic Chemicals which was redistilled (b.p. 117–118°) before use. The amount of ethylenediamine employed is sufficient to provide three equivalents of diamine for each mole of chromium(II) and to neutralize any acid remaining in the chromium(II) perchlorate solution.

3. The submitters employed an aqueous $1.64M$ solution of chromium(II) perchlorate which was prepared by stirring a mixture of 5.7 g. of pure chromium metal pellets (United Mineral and Chemical Corporation, 129 Hudson St., New York, 10013) with 60 ml. of aqueous 20% perchloric acid under a nitrogen atmosphere at 30° for 12 hours.[2,3] The rate of dissolution of the chromium metal is increased if the metal is washed successively with concentrated hydrochloric acid and with water just before it is added to the perchloric acid. The checkers employed aqueous $0.519M$ chromium(II) perchlorate solution which was prepared in a comparable manner with chromium metal obtained from the Mining and Metals Division, Union Carbide Corporation. The deep blue solution of chromium(II) perchlorate is transferred to a storage vessel with a

siphon or a hypodermic syringe and the solution is stored under a nitrogen atmosphere in a vessel fitted with a rubber septum. Provided this solution is protected from oxygen, it is stable for long periods of time; aliquots for standardization or reaction are conveniently removed with a hypodermic syringe. The solution is standardized by adding 5.00-ml. aliquots to excess aqueous $1M$ iron(III) chloride followed by titration of the iron(II) ion produced with standard cerium(IV) sulfate solution and phenanthroline as an indicator.[4]

The submitters had recommended use of only slightly more [2.3 moles of chromium(II) complex per mole of halide] than the stoichiometric amount of chromium(II) complex in this reduction. However, because these concentrations of reagents lead to a very slow reaction rate in the last 5–10% of the reduction (Note 6), the checkers found it more convenient to employ excess reducing agent [3.8 moles of chromium(II) complex per mole of halide].

4. 1-Bromonaphthalene, m.p. 2–4°, obtained either from the Aldrich Chemical Company, Inc., or from Matheson Coleman, and Bell was used without further purification.

5. A slow stream of nitrogen was passed through the dimethylformamide for 30 minutes to remove any dissolved oxygen.

6. The progress of this reaction may be followed by quenching aliquots of the reaction solution in acidic aqueous ammonium sulfate followed by extraction with ether and analysis of the ethereal extract by gas chromatography. With 1.2-m. gas chromatography column packed with silicone fluid, No. 710, on Chromosorb P and heated to 215°, the retention times of naphthalene and 1-bromonaphthalene were 1.9 minutes and 6.7 minutes, respectively. The submitters employed a 30-cm. gas chromatography column packed with Porpak P for this analysis.

Since the presence of even 5–10% of unchanged 1-bromonaphthalene makes purification of the naphthalene difficult, it is important that the reduction be complete before the product is isolated. With reaction conditions described in this preparation [0.100 mole of ethylenediamine, 0.0080 mole of 1-bromonaphthalene, 60 ml. (0.031 mole) of aqueous $0.519M$

chromium(II) perchlorate, and 80 ml. of dimethylformamide], the checkers found that reduction was usually complete in less than 15 minutes. Under the conditions [0.032 mole of ethylenediamine, 0.0040 mole of 1-bromonaphthalene, 5.5 ml. (0.0090 mole) of aqueous 1.64M chromium(II) perchlorate, and 40 ml. of dimethylformamide] originally suggested by the submitters, a reduction time of approximately 3 hours was required for complete reduction.

3. Discussion

1-Bromonaphthalene has been reduced to naphthalene in good yield by hydrogenation over Raney nickel in methanolic potassium hydroxide,[5] by triphenyltin hydride in benzene,[6] by magnesium in isopropyl alcohol,[7] by sodium hydrazide and hydrazine in ether,[8] and by copper(I) acetate in pyridine.[9]

The present procedure illustrates the ease of reduction of aryl, vinyl, and primary alkyl halides to the corresponding hydrocarbons with the chromium(II) en reagent.[10] This reagent will also convert epoxides and aliphatic halides with good leaving groups in the β-position to olefins.[10] Although the reduction of alkyl halides with this en complex is chemically similar to reductions with solutions of other chromium(II) salts in aqueous dimethylformamide,[4,11] the en complexes of the chromium(II) ion are more reactive than the aquated chromium(II) ion.[10] The checkers have found that the potential measured between platinum and calomel electrodes in a solution of chromium(II) perchlorate in aqueous dimethylformamide is increased by the addition of ethylenediamine until three equivalents of the diamine have been added. However, presumably at least one of the six coordination sites on the chromium(II) ion must be vacant at the time reduction of a halide occurs to permit transfer of the halogen atom from the substrate to the chromium ion.[10]

Chromium(II) perchlorate is the salt of choice for preparing the en complex in dimethylformamide solution. At comparable concentration levels chromium(II) sulfate[4] is insoluble, and the chromium(II) chloride[2,11] is only partially soluble in the reaction solution.

1. Department of Nematology, University of California, Riverside, California 92502.
2. H. Lux and G. Illmann, *Chem. Ber.*, **91**, 2143 (1958).
3. D. G. Holah and J. P. Fackler, *Inorg. Syn.*, **10**, 29 (1967).
4. A. Zurqiyah and C. E. Castro, *Org. Syn.*, **49**, 98 (1969).
5. H. Kämmerer, L. Horner, and H. Beck, *Chem. Ber.*, **91**, 1376 (1958).
6. L. A. Rothman and E. I. Becker, *J. Org. Chem.*, **25**, 2203 (1960).
7. D. Bryce-Smith, B. J. Wakefield, and E. T. Blues, *Proc. Chem. Soc. (London)*, 219 (1963).
8. T. Kaufmann, H. Henkler, and H. Zengel, *Angew. Chem.*, **74**, 248 (1962).
9. R. G. R. Bacon and H. A. O. Hill, *J. Chem. Soc.*, 1112 (1964).
10. J. K. Kochi, D. M. Singleton, and L. J. Andrews, *Tetrahedron*, **24**, 3503 (1968).
11. J. R. Hanson and E. Premuzic, *Angew. Chem., Int. Ed. Engl.*, **7**, 247 (1968).

MACROCYCLIC POLYETHERS: DIBENZO-18-CROWN-6 POLYETHER AND DICYCLOHEXYL-18-CROWN-6 POLYETHER

(Dibenzo [b, k] [1, 4, 7, 10, 13, 16,] hexaoxacyclooctadecin, 6, 7, 9, 10, 17, 18, 20, 21-octahydro- and dibenzo [b, k] [1, 4, 7, 10, 13,16] hexaoxacyclooctadecin, eicosahydro-)

Submitted by Charles J. Pedersen[1]
Checked by Edith Feng and Herbert O. House

1. Procedure

Caution! The subsequently described macrocyclic polyethers are toxic (Note 1) and should be handled with care.

A. *Dibenzo-18-crown-6 polyether.* A dry 5-l. three-necked flask is fitted with a reflux condenser, a 500-ml. pressure-equalizing dropping funnel, a thermometer, and a mechanical stirrer. An inlet tube at the top of the reflux condenser is used to maintain a static nitrogen atmosphere in the reaction vessel throughout the reaction. The flask is charged with 330 g. (3.00 moles) of catechol (Note 2) and 2 l. of commercial 1-butanol and then stirring is started and 122 g. (3.05 moles) of sodium hydroxide pellets is added. The mixture is heated rapidly to reflux (about 115°) and then a solution of 222 g. (1.55 moles) of bis(2-chloroethyl) ether (Note 3) in 150 ml. of 1-butanol is added, dropwise with continuous stirring and heating, over a 2-hour period. After the resulting mixture has been refluxed with stirring for an additional hour, it is cooled to 90° and an additional 122 g. (3.05 moles) of sodium hydroxide pellets is added. The mixture is refluxed with stirring for 30 minutes and then a solution of 222 g. (1.55 moles) of bis(2-chloroethyl) ether (Note 3) in 150 ml. of 1-butanol is added, dropwise with stirring and heating, over a period of 2 hours. The final reaction mixture is refluxed, with stirring, for 16 hours (Note 4) and then acidified by the dropwise addition of 21 ml. of aqueous concentrated hydrochloric acid. The reflux condenser is replaced with a distillation head and approximately 700 ml. of 1-butanol is distilled from the mixture. As the distillation is continued, water is added to the flask from the dropping funnel at a sufficient rate to maintain a constant volume in the reaction flask. This distillation is continued until the temperature of the distilling vapor exceeds 99° (Note 5), and then the resulting slurry is cooled to 30–40°, diluted with 500 ml. of acetone, stirred to coagulate the precipitate, and filtered with suction. The residual crude product is stirred with 2 l. of water, filtered with suction, stirred with 1 l. of acetone, and again filtered with suction. The residual product is washed with an additional 500 ml. of acetone and then sucked dry.

This product, tan fibrous crystals melting at 161–162°, amounts to 221–260 g. (39–48%) and is sufficiently pure for use in the next part of this preparation. The product may be purified further by recrystallization from benzene. The dibenzo-18-crown-6 polyether separates as white fibrous needles melting at 162.5–163.5° (Note 6).

B. *Dicyclohexyl-18-crown-6 polyether.* A 1-l. stainless steel autoclave is charged with a mixture of 125 g. (0.347 mole) of dibenzo-18-crown-6 polyether, 500 ml. of redistilled 1-butanol (Note 7), and 12.5 g. of 5% ruthenium-on-alumina catalyst (Note 8). After the autoclave has been closed, it is flushed with nitrogen and filled with hydrogen. The mixture is hydrogenated at 100° and a hydrogen pressure of about 70 atm. (1000 p.s.i.) until the theoretical amount of hydrogen (2.08 moles) has been absorbed. The autoclave is cooled to room temperature and then vented and the reaction mixture is filtered to remove the catalyst (Note 9). The filtrate is concentrated under reduced pressure at 90–100° with a rotary evaporator (Note 10). The residual crude product solidifies on standing (Note 11). To remove hydroxylic impurities, a solution of the crude product (about 130 g.) in 400 ml. of n-heptane is filtered through a 7-cm. by 20-cm. column of acid-washed alumina (80–100 mesh, activity I–II), and the column is eluted with additional n-heptane until the eluate exhibits hydroxyl absorption in the 3-μ region of the infrared. The solvent is removed from the combined eluates with a rotary evaporator to leave 75–89 g. (58–69%) of mixture of diastereoisomeric dicyclohexyl-18-crown-6 polyethers as white prisms melting within the range 38–54° (Note 12). This mixture of stereoisomers may be used to prepare complexes with various metal salts (Notes 13 and 14).

2. Notes

1. Dicyclohexyl-18-crown-6 polyether possesses unusual physiological properties which require care in its handling.[4] It is likely that other cyclic polyethers with similar complexing power are also toxic, and should be handled with equal care.

Oral toxicity. The approximate lethal dose of the dicyclohexyl-18-crown-6 polyether for ingestion by rats was 300 mg./kg. In a 10-day subacute oral test, the compound did not exhibit any cumulative oral toxicity when administered to male rats at a dose level of 60 mg./kg./day. It should be noted that dosage at the approximate lethal dose level caused death in 11 minutes, but that a dose of 200 mg./kg. was not lethal in 14 days.

Eye irritation. This dicyclohexyl polyether produced some generalized corneal injury, some iritic injury, and conjunctivitis when introduced as a 10% solution in propylene glycol. Although tests are not complete, there may be permanent injury to the eye even if the eye is washed after exposure.

Skin absorption. Dicyclohexyl-18-crown-6 polyether is very readily absorbed through the skin of test animals. It caused fatality when absorbed at the level of 130 mg./kg.

Skin irritation. Primary skin irritation tests run on this polyether indicate the material should be considered a very irritating substance.

2. Catechol of satisfactory purity may be purchased from Eastman Organic Chemicals or from Aldrich Chemical Company, Inc.

3. Bis(2-chloroethyl) ether may be obtained from Eastman Organic Chemicals. The checkers redistilled this material (b.p. 175–177°) before use.

4. A shorter period of refluxing may be sufficient.

5. The bulk of the material, a 1-butanol-water azeotrope, distils at 92°.

6. The product has ultraviolet maxima (CH_3OH solution) at 223 mμ (ϵ 17,500) and 275 mμ (ϵ 5500) with n.m.r. peaks ($CDCl_3$ solution) at 6.8–7.0 (8H multiplet, aryl CH) and 3.8–4.3 p.p.m. (16H multiplet, CH_2—O). The mass spectrum exhibits the following abundant peaks: m/e (rel. int.), 360 (M^+, 29), 137 (29), 136 (74), 121 (100), 109 (23), 80 (31), 52 (21), 45 (27), and 43 (34).

7. It is advisable to use redistilled solvent to avoid the presence of catalyst poisons.

8. The 5% ruthenium-on-alumina catalyst is available from Engelhard Industries.

9. Since the catalyst, saturated with hydrogen, is pyrophoric, it should be kept wet with water after the filtration has been completed.

10. Since the product, a polyether, is apt to be oxidized by air, especially at elevated temperatures in the molten state, the product should be stored under a nitrogen atmosphere.

11. This residue is a mixture of stereoisomeric dicyclohexyl-18-crown-6 polyethers which may be contaminated with some unchanged dibenzo-18-crown-6 polyether and with alcohols arising from hydrogenolysis of the polyether ring. The submitter reports that this residue is sufficiently pure for many purposes such as the preparation of complexes with potassium hydroxide which are soluble in aromatic hydrocarbons.

12. The submitter reports that the two major diastereoisomers present, designated isomer A, m.p. 61–62° and isomer B, as one of two crystalline forms, m.p. 69–70° or m.p. 83–84°, may be separated by chromatography on alumina.[2] An x-ray crystal structure determination for the complex of barium thiocyanate with isomer A of dicyclohexyl-18-crown-6 polyether has shown this polyether to have the *cis-syn-cis* stereochemistry.[3a] An x-ray crystal structure determination for the complex of sodium bromide with isomer B has shown this isomer to have the *cis-anti-cis* stereochemistry.[3b] The mixture of isomers A and B has negligible ultraviolet absorption (95% EtOH solution) and exhibits n.m.r. multiplets (C_6D_6 solution) in the regions 3.3–4.0 (20H multiplet, O—CH) and 0.9–2.2 p.p.m. (16H multiplet, aliphatic CH). The mass spectrum of the mixture exhibits the following relatively abundant peaks: m/e (rel. int.), 372 (M^+, 2), 187 (35), 143 (100), 141 (47), 99 (92), 98 (46), 97 (41), 89 (66), 87 (41), 83 (45), 82 (55), 81 (99), 73 (77), 72 (46), 69 (58), 67 (42), 57 (50), 55 (58), 45 (77), 43 (61), and 41 (58). Although the infrared (CCl_4 solution) and n.m.r. (C_6D_6 solution, 100 mHz.) spectra of the pure isomers A and B differ slightly from one another, the checkers were unable to use these spectra to determine quantitatively the composition

of mixtures of the two isomers. The most notable difference in these spectra is the shape of the n.m.r. multiplet in the region 3.3–4.0 p.p.m.; this multiplet is considerably broader in isomer A than in isomer B allowing a qualitative estimate of the purity of each isomer.

13. The submitter prepared a toluene solution of the complex of potassium hydroxide with dicyclohexyl-18-crown-6 polyether by the following procedure. A mixture of 14.9 g. (0.0402 mole) of dicyclohexyl-18-crown-6 polyether (mixture of isomers) and 2.64 g. (0.0400 mole) of 85% potassium hydroxide was dissolved in 50 ml. of methanol with gentle warming on a steam bath. The solution was diluted with 100 ml. of toluene and then concentrated with a rotary evaporator to a volume of 50 ml. An additional 100 ml. of toluene was added, and the solution was again concentrated to a volume of 50 ml. This solution was diluted with toluene to a volume of 100 ml., 1 g. of decolorizing charcoal was added, and the mixture was allowed to stand overnight under a nitrogen atmosphere. After gravity filtration, a clear toluene solution of the complex was obtained. Titration with standard aqueous hydrochloric acid indicated the solution to be approximately $0.3 M$ in base. This solution, which must be protected from atmospheric moisture and carbon dioxide, has been used for the saponification of sterically hindered esters.[4]

14. The checkers prepared a crystalline complex of potassium acetate with isomer B of dicyclohexyl-18-crown-6 polyether by the following procedure. To a stirred solution of 15.0 g. (0.0404 mole) of dicyclohexyl-18-crown-6 polyether (mixture of isomers) in 50 ml. of methanol was added a solution of 5.88 g. (0.0600 mole) of anhydrous potassium acetate (dried at 100° under reduced pressure) in 35 ml. of methanol. The resulting solution was concentrated under reduced pressure with a rotary evaporator, and the residual white solid was extracted with 35 ml. of boiling methylene chloride. The resulting mixture was filtered and the filtrate was cooled in a dry ice-acetone bath and slowly diluted with petroleum ether (b.p. 30–60°, approximately 200 ml. was required) to initiate crystallization. The

resulting suspension of the crystalline complex was allowed to warm to room temperature and was filtered with suction. Recrystallization of this complex from a methylene chloride-petroleum ether (b.p. 30–60°) mixture separated 4.21–4.35 g. (22–23%) of the complex of potassium acetate with isomer B of dicyclohexyl-18-crown-6 polyether as white needles, m.p. 165–250° dec. This complex has infrared absorption (CH_2Cl_2 solution) at 1570 and 1385 cm.$^{-1}$ (COO^-) with n.m.r. absorption ($CDCl_3$ solution) at 3.3–4.0 (20H multiplet, O—CH), 1.95 (3H singlet, CH_3CO), and 1.0–2.1 p.p.m. (16H multiplet, aliphatic CH). A 4.21-g. sample of this complex was partitioned between 75 ml. of water and three 25-ml. portions of ether. The combined ether solutions were dried over anhydrous magnesium sulfate and then concentrated under reduced pressure to leave 1.82 g. of isomer B of dicyclohexyl-18-crown-6 polyether as white prisms, m.p. 68–69°.

3. Discussion

The preparation of dibenzo-18-crown-6 polyether directly from catechol and bis(2-chloroethyl) ether has been reported previously.[4] The present procedure is an improvement of this method. Although dibenzo-18-crown-6 polyether can be obtained in 80% yield from bis-[2-(o-hydroxyphenoxy)-ethyl] ether and bis(2-chloroethyl) ether, the former intermediate has to be synthesized by a method involving several steps. One of the hydroxyl groups of catechol must be protected against alkali by reaction with a molecule of dihydropyran or chloromethylmethyl ether. Then the intermediate is treated with bis(2-chloroethyl) ether in the presence of alkali and, finally, converted into the desired intermediate by acid hydrolysis.[4] The yield of bis[2-(o-hydroxyphenoxy)-ethyl] ether was less than 40% so that the overall yield of dibenzo-18-crown-6 polyether never approached 39–48%, the yield of the present direct method.

Dibenzo-24-crown-8 and dibenzo-30-crown-10 polyethers can be prepared by this method with the substitution of the appropriate ω,ω'-dichloropolyether for bis(2-chloroethyl) ether. However, dibenzo-12-crown-4 and macrocyclic polyethers containing two or more benzo groups and an uneven number of oxygen atoms have to be prepared by the alternative method mentioned above using the intermediate catechol monoethers. Macrocyclic polyethers containing one benzo group can be synthesized by the direct reaction between one molecule of catechol and one molecule of ω,ω'-dichloropolyethers in the presence of alkali. Certain substituted crown compounds can be obtained by using catechol derivatives, such as 4-(tertiary-butyl)-catechol and 4-chlorocatechol, which do not give side reactions in the presence of alkali.

It is unusal to form a ring of eighteen atoms in a single operation by the reaction of two molecules of catechol with two molecules of bis(2-chloroethyl) ether. It seems possible that the ring-closure step is facilitated by the presence of sodium ion which is solvated by the intermediate acyclic polyether. Some experiments appear to support this hypothesis. The yields of dibenzo-18-crown-6 polyether are higher when it is prepared with sodium or potassium hydroxide than when lithium or tetramethylammonium hydroxide is used. Lithium and quaternary ammonium ions are not strongly complexed by the polyethers. Furthermore, the best ligands for alkali metal cations, polyethers containing rings of 15 to 24 atoms including 5 to 8 oxygen atoms, are formed in higher yields than smaller or larger rings, or rings of equal sizes with only 4 oxygen atoms.

The physical properties of many macrocyclic polyethers and their salt complexes have been already described.[4,5] Dibenzo-18-crown-6 polyether is useful for the preparation of sharp-melting salt complexes. Dicyclohexyl-18-crown-6 polyether has the convenient property of solubilizing sodium and potassium salts in aprotic solvents, as exemplified by the formation of a toluene solution of the potassium hydroxide complex (Note 13). Crystals of potassium permanganate, potassium t-butoxide, and potassium palladium(II) tetrachloride ($PdCl_2 + KCl$) can be made to dissolve in liquid aromatic hydrocarbons merely by adding dicyclohexyl-18-crown-6 polyether.[4] The solubilizing power of the saturated macrocyclic polyethers permits ionic reactions to occur in aprotic media. It is expected that this property will find practical use in catalysis, enhancement of chemical reactivity, separation and recovery of salts, electrochemistry, and in analytical chemistry. There are some limitations. Although salts with high lattice energy, such as fluorides, nitrates, sulfates and carbonates, form complexes with macrocyclic polyethers in alcoholic solvents as readily as more polarizable (softer) salts, their complexes cannot be isolated in the solid state because one or the other uncomplexed component precipitates on concentrating the solutions. For the same reason, these salts cannot be rendered soluble in aprotic solvents by the polyethers.

1. Contribution No. 244 from Elastomer Chemicals Department, Research Division, Experimental Station, E. I. duPont de Nemours and Co., Wilmington, Delaware 19898.
2. For further details, see H. K. Frensdorff, *J. Amer. Chem. Soc.*, **93**, 4684 (1971).
3. (a) N. K. Dalley, D. E. Smith, R. M. Izatt, and J. J. Christensen, *Chem. Commun.*, 90 (1972); (b) D. E. Fenton, M. Mercer, and M. R. Truter, *Biochem. Biophys. Res. Comm.*, in press.
4. C. J. Pedersen, *J. Amer. Chem. Soc.*, **89**, 7017 (1967); **92**, 386 (1970).
5. For reviews see (a) J. J. Christensen, J. O. Hill, and R. M. Izatt, *Science*, **174**, 459 (1971); (b) C. J. Pedersen and H. K. Frensdorff, *Angew. Chem., Int. Ed. Engl.*, **11**, 16 (1972).

METALATION OF 2-METHYLPYRIDINE DERIVATIVES: ETHYL 6-METHYLPYRIDINE-2-ACETATE

(6-Methylpyridine-2-acetic acid, ethyl ester)

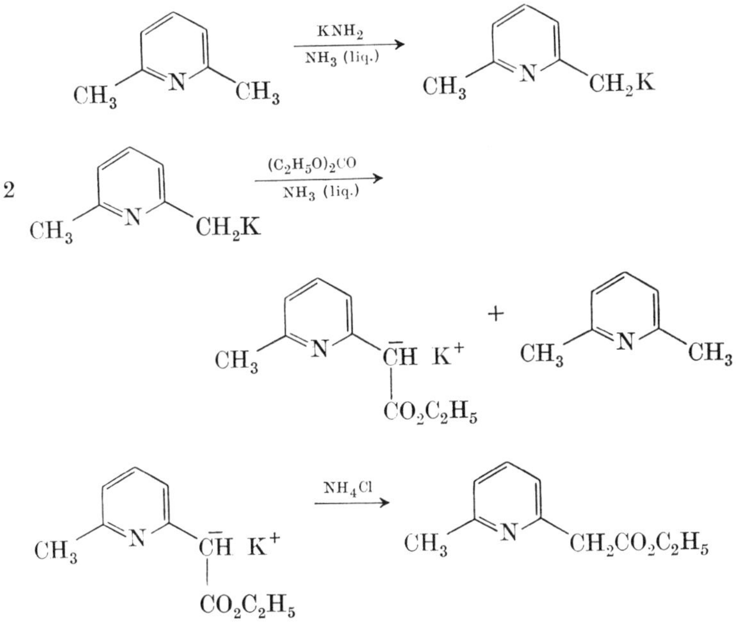

$$\text{CH}_3\text{-pyridine-CH}_3 \xrightarrow[\text{NH}_3 \text{ (liq.)}]{\text{KNH}_2} \text{CH}_3\text{-pyridine-CH}_2\text{K}$$

$$2\ \text{CH}_3\text{-pyridine-CH}_2\text{K} \xrightarrow[\text{NH}_3 \text{ (liq.)}]{(\text{C}_2\text{H}_5\text{O})_2\text{CO}} \text{CH}_3\text{-pyridine-}\bar{\text{C}}\text{H K}^+ \underset{\text{CO}_2\text{C}_2\text{H}_5}{|} + \text{CH}_3\text{-pyridine-CH}_3$$

$$\text{CH}_3\text{-pyridine-}\bar{\text{C}}\text{H K}^+ \underset{\text{CO}_2\text{C}_2\text{H}_5}{|} \xrightarrow{\text{NH}_4\text{Cl}} \text{CH}_3\text{-pyridine-CH}_2\text{CO}_2\text{C}_2\text{H}_5$$

Submitted by WILLIAM G. KOFRON and LEONA M. BACLAWSKI[1]
Checked by R. E. IRELAND and R. A. FARR

1. Procedure

Caution! This preparation should be carried out in a good hood to avoid exposure to ammonia.

A 1-l. three-necked flask is fitted with a dry ice-acetone condenser, a glass stirrer, and a glass stopper (Note 1). Potassium amide is prepared in 400 ml. of liquid ammonia from 8.0 g. (0.20 g.-atom) of potassium metal (Note 2). The glass stopper is replaced with an addition funnel containing 32.1 g. (0.300 mole) (Note 3) of 2,6-lutidine dissolved in about 20 ml. of anhydrous ether. The lutidine solution is added to the amide solution and the funnel is rinsed with a little ether, which is also added. The resulting orange solution of potassiolutidine is stirred for 30 minutes, and then cooled in a dry ice-acetone bath (Note 3). As rapidly as possible, 11.8 g. (0.100 mole) of freshly distilled diethyl carbonate is added, and the cooling bath is removed. The color changes to green. After 5 minutes the reaction mixture is neutralized by the addition of 10.7 g. (0.200 mole) of ammonium chloride. The green color is discharged, and the final mixture is gray. The condenser is removed, and the ammonia is allowed to evaporate (Note 4). The residue is stirred with 500 ml. of ether and filtered. The residual solid is extracted with an additional 100 ml. of ether, and the ethereal extracts are combined and concentrated with a rotary evaporator. The residual oil is distilled using a modified Claisen flask. Lutidine (22 g., 69%) is collected at 46–56° (10 mm.) and 10.7–13.4 g. (59–75%) of the ester is collected at 87° (0.7 mm.). The ester is a bright yellow liquid, n^{25} D 1.4995, d_4^{20} 1.0608 (Note 5).

2. Notes

1. A Teflon stirrer should not be used since Teflon is attacked by alkali metals, by metal amides, and by carbanions.

2. The preparation of potassium amide is described in a previous volume.[2] *Caution! Potassium may form an explosive red or orange peroxide coating. Potassium is a silver-gray metal with a blue-violet cast*, and any potassium showing an orange or red color, or with an appreciable oxide coating, should be considered extremely hazardous.[3]

3. The use of an extra mole of lutidine and rapid addition of diethyl carbonate both decrease formation of urethane. The use of the dry ice bath to cool the reaction mixture permits rapid addition of diethyl carbonate without excessive foaming. If urethane is formed from the diethyl carbonate and ammonia, the yield of product is decreased and the distillation is difficult.

4. A steam bath or hot air gun may be used with care to speed up evaporation of the ammonia.

5. The ester has been reported[4] to boil at 132° (18 mm.). It was reported as a colorless oil, giving a hydrochloride melting

at 112–115°. Ethyl pyridine-2-acetate was reported in a previous volume[5] as a light yellow liquid, b.p. 135–137° (28 mm.). The n.m.r. spectrum of the present product is in accord with the structure assigned and the hydrochloride melts at 115°.

3. Discussion

Ethyl pyridine-2-acetate[5] and ethyl 6-methylpyridine-2-acetate have previously been prepared by carboxylation of the lithio derivatives of α-picoline and lutidine, respectively. Use of ethyl carbonate to acylate the organometallic derivative avoids the intermediacy of the (unstable) carboxylic acid, and the yields are better. In the present procedure potassium amide is used as the metalating agent; the submitters report that the same esters may be formed by metalation with sodium amide (43% yield) or with n-butyllithium (39% yield). The latter conditions also yield an appreciable amount of the acid (which decarboxylates).

1. Department of Chemistry, University of Akron, Akron, Ohio.
2. S. Boatman, T. M. Harris, and C. R. Hauser, *Org. Syn.*, **48**, 40 (1963). See also L. F. Fieser and M. Fieser, "Reagents for Organic Synthesis," John Wiley & Sons, New York, 1967, p. 907.
3. See J. F. Short, *Chem. Ind.* (*London*), 2132 (1964) and references therein; see also D. P. Mellor, *Chem. Ind.* (*London*), 723 (1965) and M. S. Bil, *Chem. Ind.* (*London*), 812 (1965).
4. V. Boekelheide and W. G. Gall, *J. Org. Chem.*, **19**, 499 (1954).
5. R. B. Woodward and E. C. Kornfeld, *Org. Syn.*, Coll. Vol. **3**, 413 (1955).

2-METHYL-2-NITROSOPROPANE AND ITS DIMER

$$t\text{-}C_4H_9\text{—}NH_2 \xrightarrow[H_2O,55°]{KMnO_4} t\text{-}C_4H_9\text{—}NO_2$$

$$\xrightarrow[H_2O,(C_2H_5)_2O]{Al\text{-}Hg} t\text{-}C_4H_9\text{—}NHOH \xrightarrow[H_2O,-20\text{ to }25°]{NaOBr}$$

$$t\text{-}C_4H_9\text{—}N\text{=}O \rightleftharpoons \begin{array}{c} t\text{-}C_4H_9 \diagdown \quad \diagup O^- \\ \overset{+}{N}\text{=}\overset{+}{N} \\ \diagup \quad \diagdown \\ {}^-O \quad \quad C_4H_9\text{-}t \end{array}$$

Submitted by A. CALDER, A. R. FORRESTER,[1] and S. P. HEPBURN
Checked by DAVID S. CRUMRINE and HERBERT O. HOUSE

1. Procedure

A. *2-Methyl-2-nitropropane.* To a well-stirred suspension of 650 g. (4.11 moles) of potassium permanganate in 3 l. of water, contained in a 5-l. three-necked flask fitted with a reflux condenser, a mechanical stirrer, a thermometer, and a 250-ml. dropping funnel, is added dropwise and with stirring over a 10-minute period, 100 g. (1.37 moles) of *t*-butylamine (Note 1). When the addition is complete, the reaction mixture is heated to 55° over a period of approximately 2 hours, and then the reaction mixture is maintained at 55° with continuous stirring for 3 hours. The dropping funnel and reflux condenser are replaced by a stopper and a still head fitted for steam distillation and the product is steam distilled from the reaction mixture (Note 2). The liquid product is separated from the denser water layer and then diluted with 250 ml. of ether and washed successively with two 50-ml. portions of aqueous $2M$ hydrochloric acid and with 50 ml. of water. After the ethereal solution has been dried over anhydrous magnesium sulfate, the solution is fractionally distilled at atmospheric pressure to remove the ether. The residual crude product (Note 3) amounts to 106–128 g. and is sufficiently pure for use in the next step. In a typical run, distillation of 124 g. of the crude product afforded 110 g. (78%) of the pure 2-methyl-2-nitrobutane as a colorless liquid, b.p. 127–128°, n^{25} D 1.3992. The material slowly solidifies on standing to a waxy solid, m.p. 25–26° (Note 4).

B. *N-t-Butylhydroxylamine. Caution! Since hydrogen may be liberated during the reduction with aluminum amalgam, the reaction should be conducted in a hood. Also, the aluminum amalgam may be pyrophoric. Consequently, it should be used immediately and not allowed to become dry.*

Aluminum foil (30 g. or 1.1 gram-atoms, thickness 0.002–0.003 cm.) is cut into strips 5 cm. by 25 cm., and each strip is rolled into a cylinder about 1 cm. in diameter. Each of the aluminum foil cylinders is amalgamated by immersing it in a solution of 8.0 g. (0.030 mole) of mercury(II) chloride in 400 ml. of water for 15 seconds. Each amalgamated cylinder is then rinsed successively in ethanol and in ether and added to a mixture of 1.5 l. of ether and 15 ml. of water (Note 5) contained in a 3-l. three-necked flask fitted with a dropping funnel, a mechanical stirrer, and two efficient reflux condensers in series. The reaction mixture is stirred vigorously and 60 g. (0.58 mole) of 2-methyl-2-nitropropane is added dropwise at such a rate that the ether refluxes briskly. The reaction usually exhibits a 5- to 7-minute induction period after which a vigorous reaction occurs and cooling with an ice bath is necessary. After addition of the nitro compound is complete, the reaction mixture is stirred for an additional 30 minutes, and then the stirrer is stopped and the gelatinous precipitate is allowed to settle. The colorless reaction solution is decanted through a glass wool plug into a 2-l. separatory funnel and washed with two 250-ml. portions of aqueous $2M$ sodium hydroxide (Note 6). The precipitate in the reaction flask is washed with two 500-ml. portions of ether and these washings are combined and washed with the aqueous sodium hydroxide solution (Note 6). The combined ethereal solutions are dried over anhydrous sodium sulfate and concentrated under reduced pressure with a rotary evaporator. The residual crystalline solid is dried under reduced pressure (10–15 mm.) at room temperature to leave 33.7–38.7 g. (65–75%) of the crude hydroxylamine product, m.p. 59–60°, which is sufficiently pure for use in the next step. The crude product may be recrystallized from pentane to separate the pure N-*t*-butylhydroxylamine as white plates, m.p. 64–65° (Note 7).

C. *2-Methyl-2-nitrosopropane.* A solution of sodium hypobromite is prepared by adding, dropwise, and with stirring over a 5-minute period, 57.5 g. (18.5 ml. or 0.360 mole) of bromine to a solution of 36.0 g. (0.900 mole) of sodium hydroxide in 225 ml. of water. The resulting yellow solution, contained in a 1-l. three-necked flask fitted with a mechanical stirrer, a thermometer, and a dry ice-acetone cooling bath, is cooled to $-20°$. A suspension of 26.7 g. (0.300 mole) of N-*t*-butylhydroxylamine in 50 ml. of water is added to the reaction flask with continuous stirring as rapidly as possible without allowing the temperature of the reaction mixture to exceed 0°. The reaction solution is again cooled to $-20°$, and then the cooling bath is removed and the mixture is stirred for 4 hours while the reaction mixture warms to room temperature. The solid product, the nitroso dimer which has separated, is collected on a sintered glass funnel and then pulverized and washed with 1 l. of water (Note 8). The residual solid is dried at room temperature under reduced pressure (10–15 mm.) to leave 19.6–22.2 g. (75–85%) of the dimer of 2-methyl-2-nitrosopropane, m.p. 80–81° (Note 9). The product is sufficiently pure to be stored (Note 10) for use as a free radical trapping reagent.

2. Notes

1. t-Butylamine, purchased from Aldrich Chemical Company, Inc. may be used without purification.

2. Approximately 1 l. of distillate needs to be collected to remove the product from the reaction mixture.

3. The principal contaminant is residual diethyl ether.

4. The purified product exhibits infrared bands (CCl_4 solution) at 1545 cm.$^{-1}$ (broad) and 1355 cm.$^{-1}$ (NO_2) with an ultraviolet maximum (95% EtOH solution) at 279 mμ (ϵ 24) and an n.m.r. singlet (CCl_4 solution) at 1.58 p.p.m. [$(CH_3)_3C$]. The mass spectrum has the following abundant fragment peaks: m/e (rel. int.), 57 (100), 41 (74), 39 (45), and 29 (57).

5. Since water is one of the reactants in this reduction, it is necessary that at least a stoichiometric quantity of water is present.

6. Since the hydroxylamine product is readily oxidized by air to the blue nitroso compound, these manipulations should be performed rapidly to minimize exposure of the product to atmospheric oxygen. Any nitroso compound formed at this stage will co-distil with the ether and is difficult to recover.

7. The product has infrared absorption (CCl_4 solution) at 3600 cm.$^{-1}$ and 3250 (broad) cm.$^{-1}$ (OH and NH) with n.m.r. singlets (CCl_4 solution) at 5.86 (2H, NH, and OH) and 1.09 p.p.m. [9H, $(CH_3)_3C$]. The mass spectrum has the following abundant peaks: m/e (rel. int.), 89 (M$^+$, 11), 74 (96), 58 (41), 57 (100), 56 (52), 42 (41), 41 (74), 39 (34), 29 (54), and 28 (39).

8. Thorough washing to remove the last traces of alkali is essential, or the nitroso dimer will decompose to volatile products on standing.

9. When the colorless nitroso dimer is dissolved in various solvents, it partially dissociates to form a blue solution which contains an equilibrium mixture of monomer and dimer. In benzene-d_6 and in carbon tetrachloride solutions, the n.m.r. spectrum of the initial solutions of dimer changes rapidly and equilibrium is established within 20–30 minutes. From n.m.r. measurements at about 40° the equilibrium mixtures in carbon tetrachloride and in benzene-d_6 contain 80–81% of the monomer;[12] the n.m.r. singlets attributable to t-butyl groups are observed at 1.24 (monomer) and 1.57 p.p.m. (dimer) in carbon tetrachloride and at 0.97 (monomer) and 1.49 p.p.m. (dimer) in benzene-d_6. The infrared spectrum of the equilibrated mixture (CCl_4 solution) exhibits absorption at 1565 cm.$^{-1}$ attributable to the $N=O$ group of the monomer; this peak is not observed in the infrared spectrum (KBr pellet) of the dimer. The mass spectrum of the product exhibits the following abundant fragment peaks: m/e (rel. int.), 72 (10), 57 (100), 56 (23), 55 (21), 42 (22), 41(97), 39 (55), 30 (49), 29 (74), and 28 (53). A water solution of the dimer initially is colorless and exhibits an ultraviolet maximum at 287 mμ (ϵ 8000). On standing the solution slowly turns blue. A solution of the dimer in ethanol, after standing for 20–30 minutes, exhibits maxima at 292 mμ (ϵ 682 dimer) and 686 mμ (ϵ 14.5 monomer).

10. The submitters report that if the product is stored at 0° in the dark, it may be kept indefinitely.

3. Discussion

The oxidation of t-butylamine to 2-methyl-2-nitropropane is an example of a procedure previously illustrated in *Org. Syn.*[2] N-t-Butylhydroxylamine has previously been prepared by acid-catalysed hydrolysis of 2-t-butyl-3-phenyloxazirane[3] and by oxidation of t-butylamine.[6] The procedure described here is based on a method mentioned briefly by Smith and co-workers.[7] 2-Methyl-2-nitrosopropane has also been prepared directly by oxidation of t-butylamine[6,12] but is usually obtained by oxidation of the hydroxylamine.[4]

2-Methyl-2-nitrosopropane is an excellent scavenger of free radicals and is now widely used in "spin trapping" experiments[8,9] (although it has certain disadvantages).[10] In this recently developed technique, a reactive radical is trapped by the nitroso compound and identified by analysis of the e.s.r. spectrum of the so-formed stable nitroxide radical. The perdeuterated derivative of 2-methyl-2-nitrosopropane has also been recommended for this purpose.[11] t-Butylhydroxylamine, an intermediate in the present procedure, may also be used to synthesize t-butylphenylnitrone which has been used as a "spin-trapping" reagent.[9] The reaction of 2-methyl-2-nitrosopropane with aryl Grignard reagents has been used to prepare N-aryl-N-t-butylhydroxylamines.[5]

1. Department of Chemistry, University of Aberdeen, Aberdeen, Scotland.
2. N. Kornblum and W. J. Jones, *Org. Syn.*, **43**, 87 (1964).
3. W. D. Emmons, *J. Amer. Chem. Soc.*, **79**, 5739 (1957).
4. W. D. Emmons, *J. Amer. Chem. Soc.*, **79**, 6522 (1957).
5. A. R. Forrester and S. P. Hepburn, *J. Chem. Soc. C*, 1277 (1970) and other papers in this series.
6. E. Bamberger and R. Seligman, *Ber.*, **36**, 685 (1903); R. J. Holman and M. J. Perkins, *J. Chem. Soc. C*, 2195 (1970).
7. P. A. S. Smith, H. R. Alul, and R. L. Baumgarten, *J. Amer. Chem. Soc.*, **86**, 1139 (1964).
8. M. J. Perkins, P. Ward, and A. Horsfield, *J. Chem. Soc. B*, 395 (1970); C. Lagercrantz and S. Forshult, *Acta Chem. Scand.*, **23**, 708 (1969).
9. O. H. Griffith and A. S. Waggoner, *Accts. Chem. Res.*, **2**, 17 (1969); E. G. Janzen, *Accts. Chem. Res.*, **2**, 279 (1969); **4**, 31 (1971); M. J. Perkins, *Chem. Soc., Spec. Publ.*, No. **24**, 97 (1970).
10. A. R. Forrester and S. P. Hepburn, *J. Chem. Soc. C*, 701 (1971).
11. R. J. Holman and M. J. Perkins, *J. Chem. Soc. C*, 2324 (1971).
12. J. C. Stowell, *J. Org. Chem.*, **36**, 3055 (1971).

OXIDATION WITH THE NITROSODISULFONATE RADICAL. I. PREPARATION AND USE OF DISODIUM NITROSODISULFONATE: TRIMETHYL-p-BENZOQUINONE

(Nitrosodisulfonate, disodium and benzoquinone, 2,3,5-trimethyl)

$$NaNO_2 + 2\ SO_2 + NaHCO_3 \longrightarrow HON(SO_3Na)_2 + CO_2$$

$$HON(SO_3Na)_2 + OH^- \xrightarrow[\text{stainless steel anode}]{-e^-} \ddot{:}\ddot{O}-\ddot{N}(SO_3Na)_2 + H_2O$$

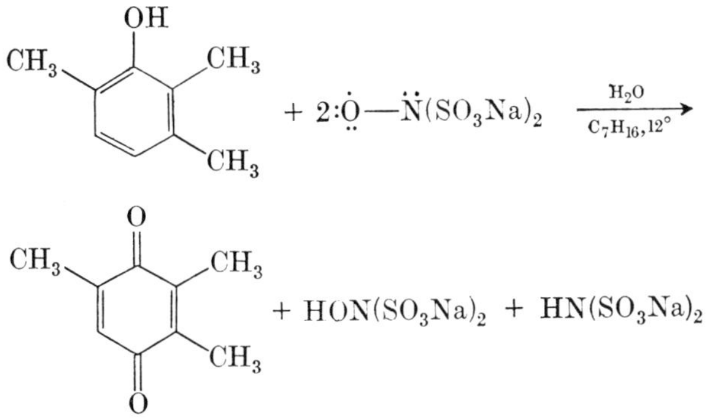

[trimethylphenol] + $2\ \ddot{:}\ddot{O}-\ddot{N}(SO_3Na)_2 \xrightarrow[C_7H_{16},\ 12°]{H_2O}$

[trimethyl-p-benzoquinone] + $HON(SO_3Na)_2 + HN(SO_3Na)_2$

Submitted by Pius A. Wehrli and Foster Pigott[1]
Checked by Don Koepsell and Herbert O. House

1. Procedure

A. *Disodium Nitrosodisulfonate.* In a 1-l. resin kettle, equipped with a mechanical stirrer, a thermometer, a gas-inlet tube suspended about 0.5 cm. above the bottom of the vessel, and an ice-cooling bath, are placed 15.0 g. (0.217 mole) of sodium nitrite (Note 1), 16.8 g. (0.200 mole) of sodium bicarbonate (Note 1), and 400 g. of ice. Sulfur dioxide (25.6 g. or 0.40 mole, Note 2) is passed into the cold, initially heterogeneous mixture with stirring over a period of 40 minutes. Near the end of the sulfur dioxide addition, the light brown color of the reaction mixture fades almost completely. The resulting colorless to pale yellow solution of disodium hydroxylaminedisulfonate (Note 3), which has an approximate pH of 4, is stirred for 10 minutes and then 59.5 g. (0.480 mole) of sodium carbonate monohydrate (Note 1) is added to give a solution of pH 11. The gas inlet tube is removed from the reaction vessel and replaced with a rectangular anode constructed from a 3.5 cm. by 4.7 cm. piece of stainless steel mesh (about 16 mesh/cm.^{2}) with a stainless steel wire as an electrical lead. The cathode is a cylindrical coil formed from a 1.5-mm. diameter by 40-cm. length of stainless steel wire which is suspended in a 5-cm. diameter by 10-cm. porous porcelain thimble filled with aqueous 10% sodium carbonate. The procelain thimble containing the cathode is suspended in the reaction vessel so that the liquid levels in the anode and cathode compartments are the same. The cathode-anode resistance of the electrolysis cell should be in the range of 5–10 ohms. While the reaction solution is continuously stirred and maintained at a temperature of 12° with an ice bath, the electrolysis is started by applying a sufficient potential (approximately 10 volts, Note 4) to the anode and cathode leads to give a cell current of 2.0 amp. As the electrolysis proceeds, the potential applied to the cell is adjusted to maintain a cell current of 2.0 amp. The formation of the nitrosodisulfonate radical is evidenced by the appearance of a deep purple color (Note 5). The electrolysis is continued with stirring and cooling until quantitative measurement of the optical density of the reaction solution (Note 5), indicates the concentration of disodium nitrosodisulfonate to be 0.42–0.47M (84–94% yield). The typical reaction time is 4 hours; the amount of electricity passed through the cell will amount to approximately 28,800 coulombs or 8 amp-hours (theoretical amount 19,300 coulombs or 5.4 amp-hours). This solution of the nitrosodisulfonate radical is removed from the anode compartment of the electrolysis cell and used directly in the next part of this preparation (Note 6).

B. *Trimethyl-p-benzoquinone.* The aqueous solution containing approximately 0.17 mole of disodium nitrosodisulfonate is placed in a 1-l. round-bottomed flask fitted with a mechanical stirrer, a thermometer, and an ice bath. A solution of 10.0g. (0.0734 mole) of 2,3,6-trimethylphenol (Note 7) in 100 ml. of heptane is added to the reaction flask and the resulting mixture is stirred vigorously for 4 hours with continuous cooling to maintain the reaction mixture below 12°. Then the yellow heptane layer is separated and the brown aqueous phase is extracted with two 100-ml. portions of heptane. The combined heptane solutions are quickly (Note 8) washed with three 50-ml. portions of cold (0–5°), aqueous $4M$ sodium hydroxide followed by two 100-ml. portions of saturated aqueous sodium chloride. The organic solution is dried over anhydrous magnesium sulfate and then concentrated at 40° under reduced pressure with a rotary evaporator. The residual crude trimethyl-p-benzoquinone amounts to 10.0–10.9 g. (91–99%) of yellow liquid which crystallizes when cooled below room temperature. Further purification may be accomplished by distillation under reduced pressure to separate 8.5–8.7 g. (77–79%) of the quinone, b.p. 53° (0.4 mm.), which crystallizes on standing as yellow needles, m.p. 28–29.5° (Note 9).

2. Notes

1. Reagent grades of these inorganic reagents were employed.
2. Sulfur dioxide was purchased from Matheson Gas Products. It is convenient to use sulfur dioxide contained in a lecture bottle so that the small cylinder can be mounted on a balance allowing continuous measurement of the weight of sulfur dioxide added.
3. The following alternative procedure may be used to prepare a solution of disodium hydroxylaminedisulfonate. Sodium nitrite (15 g., 0.217 mole) and 41.6 g. (0.40 mole) of sodium bisulfite are added to 250 g. of ice. With stirring, 22.5 ml. (0.40 mole) of acetic acid is added all at once and the mixture is stirred for 90 minutes in an ice bath. At the end of the stirring period the reaction solution is pH 5 and a potassium iodide-starch test is negative. A solution of 50 g. (0.47 mole) of sodium carbonate in water (total volume 250 ml.) is added. This buffered solution of disodium hydroxylaminedisulfonate may be used for electrolytic oxidation.
4. Since accurate control of the anode potential is not required in this oxidation, a variety of direct current sources may

be employed, provided they are able to supply continuously a current of about 2 amp. at a potential of about 12 volts. One of the simplest direct current sources is an unfiltered rectifier of the type used to charge automobile batteries.

5. In aqueous $1M$ potassium hydroxide solution, the nitrosodisulfonate radical has a maximum in the visible at 544 mμ (ϵ 14.5).

6. The submitters report that approximately half of the nitrosodisulfonate radical had decomposed after the solution was stored at 0° for 2 weeks. They report the following procedure for the isolation of Fremy's salt (dipotassium nitrosodisulfonate).

Caution! Fremy's salt may decompose spontaneously in the solid state.[2]

To the cold (12°) purple solution of disodium nitrosodisulfonate was added, dropwise and with stirring, a solution of 37.3 g. (0.5 mole) of potassium chloride in 100 ml. of water. The resulting mixture, from which the orange-yellow dipotassium nitrosodisulfonate crystals precipitated, was allowed to stand overnight in the refrigerator, and the crystals were filtered with suction and washed with 100 ml. of aqueous $1M$ potassium-hydroxide. The damp crystals weighed 55 g. A 1-g. aliquot of the wet material was dried at room temperature in a desiccator over Drierite to leave 0.76 g. of orange crystals. This corresponds to a 72% yield based on sodium nitrite. On two occasions small samples of dried material decomposed spontaneously. *It is again stressed that if the electrolyzed solution is not used directly, any isolated Fremy's salt should be stored as a slurry in 0.5M potassium carbonate at 0°.*

7. Crude 2,3,6-trimethylphenol, purchased from Aldrich Chemical Company, Inc., was purified by recrystallization from either hexane or chlorobenzene. The recrystallized phenol melted at 62–63°.

8. This extraction with aqueous sodium hydroxide to remove phenolic by-products and starting material must be performed quickly because *p*-quinones are unstable to strong bases.

9. The product has infrared absorption ($CHCl_3$ solution) at 1645 (conjugated C=O) and 1619 cm^{-1} (C=C) with ultraviolet

maxima (95% EtOH solution) at 256 mμ (ε 16,700), 341 mμ (ε 3880), and 430 mμ (shoulder, ε 30). The material has n.m.r. peaks (CDCl$_3$ solution) at 6.5–6.7 (1H multiplet, aryl CH) and 1.9–2.2 p.p.m. (9H, partially resolved multiplet, CH$_3$) with the following abundant peaks in its mass spectrum: m/e (rel. int.), 150 (100, M$^+$), 122 (24), 121 (14), 107 (30), 79 (18), 68 (20), 54 (14), 40 (18), and 39 (16).

3. Discussion

The nitrosodisulfonate salts, particularly the dipotassium salt called Fremy's salt, are useful reagents for the selective oxidation of phenols and aromatic amines to quinones (the Teuber reaction).[3,5] Dipotassium nitrosodisulfonate has been prepared by the oxidation of a hydroxylaminedisulfonate salt with potassium permanganate,[3-5] with lead dioxide,[6] or by electrolysis.[2,7] This salt is also available commercially. The present procedure illustrates the electrolytic oxidation to form an alkaline aqueous solution of the relatively soluble disodium nitrosodisulfonate. This procedure avoids a preliminary filtration which is required to remove manganese dioxide formed when potassium permanganate is used as the oxidant.[3-5]

Solutions of the nitrosodisulfonate salts are most stable in weakly alkaline solutions (pH 10) and decompose rapidly when the solution is acidic or strongly alkaline.[3] The solid dipotassium nitrosodisulfonate (Fremy's salt) has been reported to decompose spontaneously[2,3] suggesting that procedures involving the use of substantial quantities of the dry solid salt may be hazardous. In the present procedure, separation and use of the solid salt is avoided since the disodium nitrosodisulfonate is formed and used in aqueous solution. In this procedure, 2 moles of the preformed nitrosodisulfonate salt are consumed in the oxidation of one mole of the phenol to the benzoquinone derivative.[3] The submitters report that only one molar equivalent of the nitrosodisulfonate salt is required if the electrochemical oxidation is carried out in the presence of a heptane solution of the phenol.

Trimethyl-p-benzoquinone, the product of this oxidation,

has been prepared by the oxidation of 2,3,5-trimethyl-1,4-benzenediamine with iron(III) chloride[8] and by the oxidation of 2,3,5-trimethylphenol with dipotassium nitrosodisulfonate.[9]

1. Chemical Research Department, Hoffman-La Roche Inc., Nutley, N.J. 07110.
2. P. A. Wehrli and F. Pigott, *Inorg. Chem.* **9**, 2614 (1970).
3. H. Zimmer, D. C. Lankin, and S. W. Horgan, *Chem. Rev.*, **71**, 229 (1971).
4. G. Brauer, "Handbuch der Präparativen Anorganischen Chemie," Vol. 1, Ferdinand Encke Verlag, Stuttgart, 1960, p. 452.
5. H.-J. Teuber and G. Jellinek, *Chem. Ber.*, **85**, 95 (1952) and subsequent publications.
6. G. Harvey and R. G. W. Hollingshead, *Chem. Ind. (London)*, 244 (1953).
7. W. R. T. Cottrell and J. Farrar, *J. Chem. Soc. A*, 1418 (1970).
8. L. I. Smith, *J. Amer. Chem. Soc.*, **56**, 472 (1934).
9. H.-J. Teuber and W. Rau, *Chem. Ber.*, **86**, 1036 (1953).

II. USE OF DIPOTASSIUM NITROSODISULFONATE (FREMY'S SALT): 4,5-DIMETHYL-1,2-BENZOQUINONE

(o-Benzoquinone, 4,5-dimethyl-)

$$\text{3,4-dimethylphenol} + 2\,:\!\ddot{\text{O}}\!-\!\ddot{\text{N}}(\text{SO}_3\text{K})_2 \xrightarrow[(\text{C}_2\text{H}_5)_2\text{O},\,25°]{\text{H}_2\text{O}}$$

$$\text{4,5-dimethyl-}o\text{-benzoquinone} + \text{HON}(\text{SO}_3\text{K})_2 + \text{HN}(\text{SO}_3\text{K})_2$$

Submitted by H.-J. TEUBER[1]
Checked by P. A. WEHRLI, F. PIGOTT, and A. BROSSI

1. Procedure

A solution of 15 g. of sodium dihydrogen phosphate (Note 1) in 5 l. of distilled water is placed in a 6-l. separatory funnel.

To this solution is added 90 g. (0.33 mole) of potassium nitrosodisulfonate (Fremy's salt) (Note 2). The mixture is shaken to dissolve the inorganic radical. A solution of 16 g. (0.131 mole) of 3,4-dimethylphenol (Note 3) in 350 ml. of ether is added quickly to the purple solution. The mixture is shaken vigorously for 20 minutes (Note 4). The color of the solution changes to red-brown. The o-quinone thus formed is subsequently extracted in three portions with a total of 1.2 l. of chloroform. The combined organic layers are dried over anhydrous sodium sulfate (Note 5), filtered, and evaporated under reduced pressure at 20–23° (Note 6). The residual, somewhat oily red-brown crystals are slurried twice with 15 ml.-portions of ice-cold ether and collected on a filter. The remaining dark red crystals, after air drying, weigh 8.7–8.9 g. (49–50%), m.p. 105–107° (Note 7).

2. Notes

1. Monobasic sodium phosphate, $NaH_2PO_4 \cdot H_2O$, obtained from Merck & Co., Inc., was used. This buffer was found to be satisfactory for this reaction.

2. Fremy's salt may be purchased from Aldrich Chemical Company, Inc. or from Matheson Coleman and Bell. The Fremy's salt used by the checker was prepared electrolytically.[2]

3. 3,4-Dimethylphenol was obtained from Eastman Organic Chemicals; the melting point of this material was 63–65°.

4. An efficient stirrer may be substituted for the shaking.

5. The drying was accomplished in about 5 minutes.

6. Higher temperatures may accelerate dimerization of the product.

7. The product is reported to melt at 102°.[3] This material has n.m.r. peaks ($CDCl_3$ solution) at 2.14 and 6.19 p.p.m. with relative intensities of 3:1. The infrared spectrum ($CHCl_3$ solution) shows the strongest absorption at 1670 cm^{-1} accompanied, among others, by four more bands at 1390, 1280, 1005, and 835 cm^{-1}. The product has ultraviolet maxima ($CHCl_3$ solution) at 260 mμ (ϵ 2600), 400 mμ (ϵ 1120), and 572 mμ (ϵ 288). It is reported that the material undergoes slow Diels-Alder dimerization.[4]

3. Discussion

o-Quinones exemplify a very important and reactive class of compounds for general organic synthesis. In the past they have been prepared from catechol derivatives by silver oxide dehydrogenation.[3] The unique oxidizing properties of Fremy's salt allow a number of readily available phenols to be converted to *o*-quinones in excellent yield.[4] The scope of this oxidation, the Teuber reaction, is the subject of numerous papers[5] which have been reviewed recently.[6]

1. H.-J. Teuber, Institut für Organische Chemie der Universität, Frankfurt/Main.
2. P. A. Wehrli and F. L. Pigott, *Inorg. Chem.* **9**, 2614 (1970).
3. R. Willstätter and F. Müller, *Ber.*, **44**, 2171 (1911).
4. H.-J. Teuber and G. Staiger, *Chem. Ber.*, **88**, 802 (1955); H.-J. Teuber, U.S. Patent 2,782, 210 (1957).
5. H.-J. Teuber and S. Benz, *Chem. Ber.*, **100**, 2918 (1967) and earlier papers.
6. H. Zimmer, D. C. Lankin and S. W. Horgan, *Chem. Rev.*, **71**, 229 (1971).

PREPARATION OF CYANO COMPOUNDS USING ALKYLALUMINUM INTERMEDIATES. I. DIETHYLALUMINUM CYANIDE

(Cyanodiethylaluminum)

$$Al(C_2H_5)_3 + HCN \xrightarrow{C_6H_6} (C_2H_5)_2AlCN + C_2H_6$$

Submitted by W. NAGATA and M. YOSHIOKA[1]
Checked by S. C. WELCH, P. BEY, and ROBERT E. IRELAND

1. Procedure

Caution! This preparation should be conducted in a well-ventilated hood, and neat triethylaluminum must be handled with great care.

A tared 500-ml. round-bottomed flask is fitted with a vacuum take-off, and the entire assembly is connected through an adaptor containing a stopcock to an inverted cylinder of triethylaluminum as shown in Figure 1. The assembly is connected to a nitrogen source (Note 1) through the vacuum take-off and with the cylinder valve closed but the stopcock open, the system is alternately evacuated and filled with nitrogen four times. With the system filled with nitrogen the cylinder valve is opened and approximately 55 ml. (45.7 g; 0.40 mole) of triethylaluminum (Note 2) is allowed to flow into the reaction flask. The cylinder valve is then closed; the system is evacuated and filled three times with nitrogen, and the adaptor stopcock

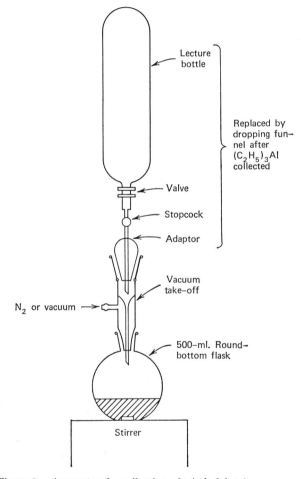

Figure 1. Apparatus for collection of triethylaluminum.

is then closed (Note 3). The reaction flask is then quickly removed, stoppered, and weighed to determine the exact amount of triethylaluminum collected. A magnetic stirring bar is added and the flask is fitted with a vacuum take-off and 250-ml. pressure-equalizing dropping funnel. The system is again placed under a nitrogen atmosphere, and the triethylaluminum is dissolved in 150 ml. of anhydrous benzene which is added through the dropping funnel. The dropping funnel is charged with 11.9 g. (0.44 mole) of hydrogen cyanide (Note 4) in 100 ml. of anhydrous benzene and then this solution is added dropwise to the solution of triethylaluminum with good magnetic stirring and cooling. Preferably, the addition is carried out at such a constant rate that the hydrogen cyanide solution is added in about 2 hours. The evolution of ethane gas becomes slow suddenly after one molar equivalent of hydrogen cyanide is added (Note 5). After the addition is complete, the reaction mixture is allowed to stir overnight (Note 6).

After this period, the dropping funnel and the vacuum take-off are replaced by the short-path distillation assembly shown in Figure 2. The system is protected by a Drierite tube and the benzene is distilled under reduced pressure (water aspirator). After the benzene is removed, the benzene-containing receiver is replaced with a clean, dry flask, and the system is connected to an efficient vacuum pump. The pressure in the system is reduced to 0.02 mm., and the flask is immersed deeply in an oil bath (Figure 2) heated to about 200°. After about 1 ml. of fluid forerun is collected, the diethylaluminum cyanide distils at 162° (0.02 mm.) (Note 7) and is collected in a tared 200-ml. receiver by heating the side arm and the adaptor with a stream of hot air or an infrared lamp (Note 8). After all the distillate is collected in the receiver (Note 9), dry nitrogen is admitted to the evacuated apparatus and the receiver is stoppered and weighed. Diethylaluminum cyanide is obtained usually as a pale yellow syrup (Note 10) in 60–80% yield (26.7–35.6 g.) (Note 11).

The stopper of the flask is quickly replaced by the nitrogen adaptor, and after placing the system under a nitrogen atmosphere, the diethylaluminum cyanide is treated with 130 ml.

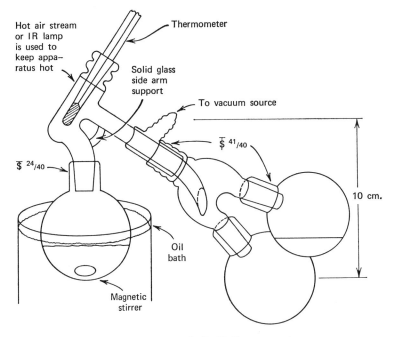

Figure 2. Short-path distillation apparatus.

of dry benzene. The resulting mixture is allowed to stand with occasional swirling under nitrogen until the syrup goes into solution (Note 12). Sufficient dry benzene is then added to make the total volume of the solution 200 ml. After thorough mixing is assured by stirring with a magnetic bar, the resulting diethylaluminum cyanide solution (13.4–17.8%; 1.2–1.6M) may be divided and stored in sealed ampoules (Note 13).

2. Notes

1. The nitrogen source described in *Org. Syn.* [Coll. Vol. **4**, 133 (1963), Figure 5] was used.

2. The exact volume of the triethylaluminum added at this point is not critical, since the exact weight is determined later. The use of a 25% solution of triethylaluminum in benzene, available from the Stauffer Chemical Company, 299 Park Avenue, New York, eliminates the tedious preparation of the triethylaluminum solution described in this procedure.

3. These precautions will ensure the removal of any adhering triethylaluminum that will flame when the apparatus is disassembled.

4. A 10% molar excess of hydrogen cyanide was employed, and the quantity added at this point was determined by the amount of triethylaluminum collected. See Note 4 in Part III of this preparation for the preparation of hydrogen cyanide.

5. The change in the rate of gas evolution is sometimes not clear, especially when the temperature of the hood is high. When the change is recognized distinctly, the addition of hydrogen cyanide solution may be stopped at this stage.

6. The reaction mixture containing about 13% ($1.2M$) of diethylaluminum cyanide and a small amount of ethylaluminum dicyanide may be used for most hydrocyanation processes without further purification. Care must be taken to have no unchanged triethylaluminum, since the submitters have observed that hydrocyanation of Δ^8-11-keto steroids with diethylaluminum cyanide is greatly retarded by the presence of a small amount of unchanged triethylaluminum.[2,5]

7. The checkers collected the diethylaluminum cyanide at 170–180° (0.35 mm.) and 167–175° (0.25 mm.).

8. Heating of the glassware above 150° with a hot air stream or infrared lamp is essential to make the viscous product run into the receiver.

9. The pot residue contains ethylaluminum dicyanide as a nonvolatile mass, most of which may be removed with a spatula and decomposed with isopropyl alcohol and then water. The flask is then washed with running water and 20% hydrochloric acid to remove the mass completely.

10. The submitters have obtained an almost colorless syrup by reaction of purified triethylaluminum with hydrogen cyanide, followed by repeated distillation.

11. The yield depends on the efficiency of the collection of the viscous distillate in the receiver.

12. It takes considerable time (5–10 hours) to dissolve the diethylaluminum cyanide. Magnetic stirring is not effective unless most of the material goes into solution.

13. Diethylaluminum cyanide dissolved in benzene, toluene,

hexane, or isopropyl ether and stored in ampoules is stable for a long period. The cyanide is not stable in tetrahydrofuran. The anhydrous benzene used in the reaction may be replaced by diethyl ether or diisopropyl ether.

3. Discussion

Formation of diethylaluminum cyanide from triethylaluminum and hydrogen cyanide was noted initially by the submitters[3] and later by Stearns,[4] but isolation and characterization of the product were first performed by the submitters.[5] An unpractical process comprising heating diethylaluminum chloride and sodium cyanide in benzene for 21 days has been reported.[6]

Diethylaluminum cyanide is a useful, potent reagent for hydrocyanation of various compounds. Features of this reagent as compared with the triethylaluminum-hydrogen cyanide reagent may be seen from the literature.[2,5,7]

1. Shionogi Research Laboratory, Shionogi & Co., Ltd., Osaka, Japan.
2. W. Nagata, "Proceedings of the Symposium on Drug Research," Montreal, Canada, June 1966, p. 188.
3. W. Nagata, M. Yoshioka, and S. Hirai, *Tetrahedron Lett.*, 461 (1962).
4. R. S. Stearns, U.S. Patent 3,078,263 (1963).
5. W. Nagata and M. Yoshioka, *Tetrahedron Lett.*, 1913 (1966); W. Nagata, M. Yoshioka, and S. Hirai, *J. Amer. Chem. Soc.*, **94**, 4635 (1972).
6. R. Ehrlich and A. R. Young, *J. Inorg. Nucl. Chem.*, **28**, 674 (1966).
7. W. Nagata and M. Yoshioka, "Proceedings of the Second International Congress on Hormonal Steroids," Excerpta Medica Foundation, Amsterdam, 1967, p. 327.

II. 1-CYANO-6-METHOXY-3,4-DIHYDRONAPHTHALENE

(6-Methoxy-3,4-dihydronaphthalene-1-carbonitrile)

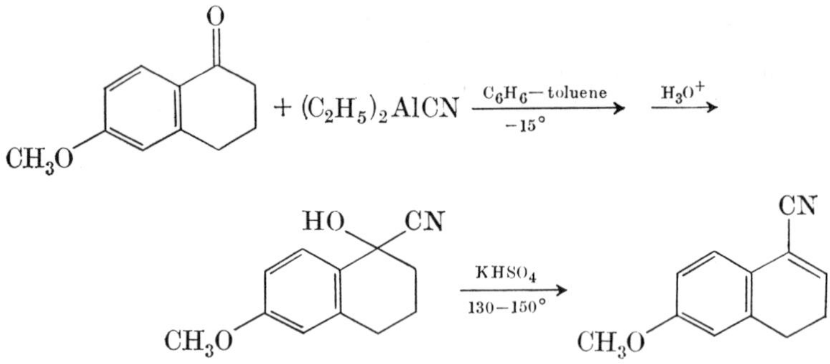

Submitted by W. Nagata, M. Yoshioka, and M. Murakami[1]
Checked by R. Wong, C. Kowalski, R. Czarny, and R. E. Ireland

1. Procedure

A 200-ml. two-necked round-bottomed flask charged with 6.15 g. (0.035 mole) of 6-methoxy-1-tetralone (Note 1) and a 100-ml. round-bottomed flask are flushed with nitrogen, and each of the flasks is fitted with an adaptor with a side arm connected to a nitrogen bubbler system and then charged with 30 ml. of anhydrous toluene. The 200-ml. flask is cooled to −20° to −25° (bath temperature) (Note 2). Into the 100-ml. flask is introduced 60 ml. (0.07 mole) of a 13% solution of diethylaluminum cyanide in benzene (Note 3) with a hypodermic syringe, and this flask is cooled with ice water. The cooled diethylaluminum cyanide solution is added to the cold solution of 6-methoxytetralone with a hypodermic syringe and the resulting mixture, after being swirled, is kept at −15° for 80 minutes under nitrogen. The stopper of the flask is replaced by a glass tube which has one end extending to the bottom of the reaction flask and the other end mounted in a neck of a 2-l. three-necked flask, equipped with an efficient stirrer and containing a cold (−70°) mixture of 250 ml. of methanol and 150

II. 1-CYANO-6-METHOXY-3,4-DIHYDRONAPHTHALENE

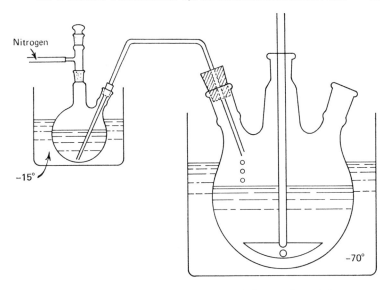

Figure 1. Apparatus for acid treatment of the reaction mixture.

ml. of concentrated hydrochloric acid as shown in Figure 1. The reaction mixture is added through the glass tube to the vigorously stirred acid mixture by applying a positive nitrogen pressure to the reaction flask (Note 4). After the bulk of the reaction mixture is added, about 50 ml. of a cold mixture of methanol and hydrochloric acid is added to the reaction flask and this mixture is transferred to the 2-l. flask in the same way described above. The stirring is continued for one hour, and the resulting mixture is poured into a mixture of 200 ml. of concentrated hydrochloric acid and 1 l. of ice water (Note 5) and extracted with three 500-ml. portions of methylene chloride. The combined organic phases are washed once with 1.5 l. of water, dried over anhydrous sodium sulfate, and evaporated from a flask containing 55 mg. of *p*-toluenesulfonic acid monohydrate (Note 6), using a rotary evaporator at a temperature below 40°.

The residue, obtained as a pale yellow oil, weighs approximately 7.4 g. and consists of 1-cyano-1-hydroxy-6-methoxytetralin and a small amount of unchanged 6-methoxy-1-tetralone. The oil is transferred to a 10-ml. Claisen flask, a small amount of a mixture of methylene chloride and ether being used to complete the transfer. Two hundred milligrams of powdered potassium bisulfate is added, and the flask is heated at 130° under reduced pressure (5 mm.) for 30 minutes. The pressure is then reduced to 0.01 mm. and the temperature is raised to about 150° to collect all the distillate [b.p. 113–117° (0.01 mm.)] in a 50-ml. flask. The viscous distillate (including material adhering to the distillation apparatus), weighs 6.0–6.2 g. and yields 4.91–5.05 g. (76–78%) of product, m.p. 50–51.5°, after two or three crystallizations from methanol. The residue from the mother liquors (1.0–1.3 g.) is adsorbed on a column of 100 times its weight of silica gel (70–325 mesh) and the column is eluted with approximately 1 l. of 40% ether in petroleum ether (b.p. 30–60°). The first 200 ml. of eluent is discarded, and then 510–550 mg. of the product is eluted in the next 250 ml. of eluent. Crystallization of this material from an ether-petroleum ether (b.p. 30–60°) mixture affords an additional 460–500 mg. (7.0–7.8%) of pure product, m.p. 50.5–51.5°. The total yield of the unsaturated nitrile is 5.41–5.51 g. (83.8–85.5%). (Note 7). The final 500 ml. fraction from chromatography contains 330–660 mg. (5.4–10.7%) of the starting material, m.p. 77–78°.

2. Notes

1. 6-Methoxy-1-tetralone will be available from K & K Laboratories, New York, although the submitters have used a material, m.p. 77–80°, produced by Osaka Yuki Gosei K. K., Nishinomiya-shi, Japan.

2. Crystals of 6-methoxy-1-tetralone may separate from the solution on cooling, but redissolve upon addition of the cooled diethylaluminum cyanide solution.

3. For preparation of diethylaluminum cyanide, see the preceding procedure. Both the submitters and checkers employed a crude reagent solution rather than a solution prepared from distilled diethylaluminum cyanide.

4. Application of the nitrogen pressure may be made conveniently by capping the outlet of the mercury bubbler.

5. The two-step decomposition is effective for preventing reconversion of the cyanohydrin into the starting ketone.

6. The cyanohydrin initially formed is unstable and readily reconverted to the starting 6-methoxy-1-tetralone on evaporation of the extracts unless slight acidity of the solution is maintained by addition of a trace amount of p-toluenesulfonic acid monohydrate. As this acid is relatively insoluble in methylene chloride, it should be added directly to the flask used for evaporation of the solvent.

7. Preferably, the product is stored in an oxygen-free atmosphere. Samples not stored in an inert atmosphere have deteriorated to dark brown masses within several months, whereas no appreciable change has been observed in a sample stored for 2 years in an ampoule filled with argon.

3. Discussion

The present method developed by the submitters[2] is the only practical process for preparation of 1-cyano-6-methoxy-3,4-dihydronaphthalene. Birch and Robinson[3] have reported that 6-methoxy-1-tetralone did not react with hydrogen cyanide or sodium acetylide.

This process presents a typical procedure applicable to preparation of cyanohydrins from ketones and aldehydes of low reactivity. 1-Cyano-6-methoxy-3,4-dihydronaphthalene is useful as an intermediate for synthesis of polycyclic compounds.

1. Shionogi Research Laboratory, Shionogi & Co., Ltd., Osaka, Japan.
2. W. Nagata and M. Yoshioka, *Tetrahedron Lett.*, 1913 (1966); W. Nagata, M. Yoshioka, and M. Murakami, *J. Amer. Chem. Soc.*, **94**, 4654 (1972).
3. A. J. Birch and R. Robinson, *J. Chem. Soc.*, 503 (1944).

III. 3β-ACETOXY-5α-CYANOCHOLESTAN-7-ONE
(3β-Hydroxy-7-oxo-5α-cholestane-5-carbonitrile 3-acetate)

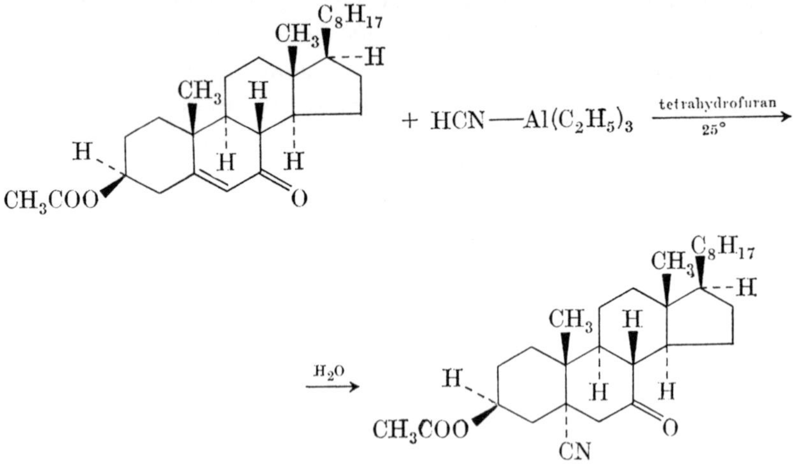

$$\text{steroid} + \text{HCN} \longrightarrow \text{Al}(C_2H_5)_3 \xrightarrow{\text{tetrahydrofuran}}_{25°}$$

$$\xrightarrow{H_2O} \text{product}$$

Submitted by W. NAGATA and M. YOSHIOKA[1]
Checked by ROBERT E. IRELAND, ROBERT CZARNY, and CONRAD J. KOWALSKI

1. Procedure

Caution! This preparation should be carried out in a good hood. Also, great care should be taken in handling neat triethylaluminum because it is pyrophoric—that is, it ignites spontaneously upon contact with air (Note 2).

A dry 50-ml. three-necked round-bottomed flask equipped with gas inlet tube for nitrogen, magnetic stirring bar, and serum stopper for the introduction of reagents is flushed with nitrogen, stoppered with a glass stopper, charged with 17 ml. of anhydrous tetrahydrofuran (Note 1) and then immersed in an ice bath. Stirring is started and 3.9 ml. (3.3 g., 0.028 mole) of triethylaluminum is introduced into the flask with a dry hypodermic syringe (Note 2, Note 3). After 5–10 minutes, 4.8 ml. of a 3.57M solution of hydrogen cyanide (0.017 mole) in anhydrous tetrahydrofuran (Note 4) is added with a dry hypodermic syringe. The stirring is continued for about 5–10 minutes.

A dry 100-ml. three-necked round-bottomed flask equipped

III. 3β-ACETOXY-5α-CYANOCHOLESTAN-7-ONE

with gas inlet tube for nitrogen, magnetic stirring bar, and serum stopper, as described above, is flushed with nitrogen. While the flask is being flushed, 2.50 g. (0.00565 mole) of 3β-acetoxy-Δ^5-cholesten-7-one (Note 5) and 0.0521 ml. (0.0029 mole) of water (Note 6) are added to the reaction flask. The flask is stoppered with a glass stopper and charged with 17 ml. of anhydrous tetrahydrofuran. After the starting material has dissolved, the cold triethylaluminum-hydrogen cyanide solution is transferred to the reaction flask using a dry hypodermic syringe. The resulting pale yellow solution is stirred at room temperature under a positive nitrogen pressure. After 3 hours, a solution of 0.044 ml. (0.0024 mole) of water in 0.87 ml. of anhydrous tetrahydrofuran is added, and the solution is allowed to stir for an additional 4 hours.

The reaction mixture is poured slowly into a vigorously stirred solution of 28 ml. (0.28 mole) of concentrated hydrochloric acid and 350 ml. of ice water placed in a 1-l. three-necked round-bottomed flask fitted with an efficient stirrer and immersed in an ice bath (Note 7, Note 8). The mixture is stirred for 20 minutes with ice cooling and extracted three times with 200-ml. portions of a 3:1 (v/v) mixture of ether and methylene chloride. The extracts are washed with three 200-ml. portions of aqueous $2M$ sodium hydroxide, two 200-ml. portions of water, and one 200-ml. portion of saturated aqueous sodium chloride, and then dried over anhydrous sodium sulfate, and evaporated under reduced pressure (Notes 9 and 10). The crystalline residue, weighing 2.70 g., is recrystallized by dissolving it in 7.5–8 ml. of hot (almost boiling) benzene and adding 25 ml. of n-pentane (distilled) to the hot solution (Note 11). 3β-Acetoxy-5α-cyanocholestan-7-one is obtained as white crystals, m.p. 192.5–193.5°; the yield is 2.27–2.41 g. (86–91%). A second crop can be obtained in 50–170 mg. yield, m.p. 188.5–190° (Note 12); the total yield is 92–93% (Notes 13 and 14).

2. Notes

1. Prior to use the tetrahydrofuran was distilled from lithium aluminum hydride into a dry flask flushed with nitrogen and sealed with a serum stopper.

2. *Caution! Triethylaluminum is pyrophoric. Use safety glasses, gloves, and an apron. Use dry sand to extinguish fires.* The submitters note that a description of the properties and handling procedures for triethylaluminum are available from the Ethyl Corporation, Louisiana. The checkers used triethylaluminum in lecture bottles from Alpha Inorganics, Inc., and suggest the handling procedure described below. Since this procedure was submitted and checked, standardized solutions of triethylaluminum in various hydrocarbon solvents, which may be substituted for pure triethylaluminum, have become available from Texas Alkyls, Inc., a Division of Stauffer Chemical Company.

A. *Checkers Handling Procedure.* Figures 1 and 2 suggest the equipment to be used, and how to assemble it. The hose end fitting is connected to the stopcock with a piece of Teflon lined tubing and fastened with copper wire. The stopcock should be well greased and held secure by means of a taut rubber band

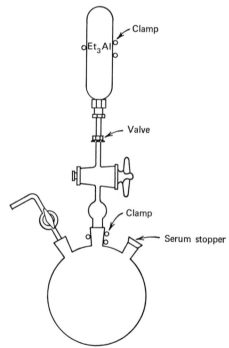

Figure 1. Apparatus for collecting triethylaluminum from a lecture bottle.

III. 3β-ACETOXY-5α-CYANOCHOLESTAN-7-ONE

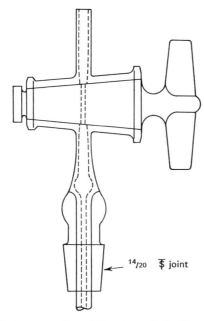

14/20 ⚡ joint

Figure 2. Detailed construction of the stopcock for the apparatus shown in Figure 1.

since it does have a tendency to pop out. The triethylaluminum may now be removed from the lecture bottle by the following procedure. A dry three-necked flask equipped with serum stopper and gas inlet tube is flushed thoroughly with nitrogen from a nitrogen bubbler. The triethylaluminum transfer apparatus is put into the open neck of the flask, the joint being previously well greased (the gas inlet tube joint should also be greased). With the stopcock of the transfer apparatus *open* and well secured as suggested above, the tank valve is opened (usually one or two full turns) with a wrench. The flow of triethylaluminum may now be adjusted with the stopcock. When one obtains as much triethylaluminum as desired, the tank valve is closed, the transfer apparatus is allowed to drain, and then the stopcock is closed. The transfer apparatus is removed from the flask which is quickly stoppered with a glass stopper while the flask is being flushed with nitrogen. This flask of triethylaluminum can be stored in this manner for many weeks since this reagent is quite stable. The transfer apparatus, which may still contain some triethylaluminum, should be carefully removed, rinsed quickly with acetone, and then cleaned with dilute hydrochloric acid.

B. *Submitters Handling Procedure.* The submitters suggest another handling procedure especially useful for removing triethylaluminum from a lecture bottle having a clogged valve outlet. Figure 3 shows the apparatus and how to transfer the material. With the cylinder clamped in upright position, remove the valve unit so that only the bottle remains. Quickly attach the adaptor to the opening of the bottle and apply a slow stream of nitrogen. Transfer the triethylaluminum into dry, nitrogen-flushed 50-ml. ampoules using a 100- or 200-ml. Luer-lock hypodermic syringe with a needle, 43 cm. long and 2 mm. in diameter. Sweep the opening of the ampoule with nitrogen during the transfer. The syringe should be slightly greased, and the ampoules should be strong with a long, thick stem so that they can be resealed. The ampoules are sealed as soon as possible. The use of rubber caps is effective for temporary protection of the ampoules from air. The material in a 50-ml. ampoule can be divided in smaller ampoules using the apparatus shown in Figure 4 which can be used also for transferring triethylaluminum from an ampoule to a reaction flask.

This procedure was not tested by the checkers. It has the advantage that one does not have to contend with a clogged lecture bottle, which in the checkers experience is best discarded. It has the disadvantage that one must handle large syringes full of triethylaluminum and make several transfers without exposing the liquid to air. Caution must be exercised with either procedure.

3. Neat triethylaluminum may be replaced by a 10–25% stock solution of it in anhydrous tetrahydrofuran with decreasing the amount of solvent in the reaction flask. The stock solution is prepared by using a graduated flask to measure the volume of the triethylaluminum and solvent added appropriately. The stock solution is very stable and not pyrophoric.

4. The submitters have prepared hydrogen cyanide as directed in *Organic Syntheses*.[2] The checkers used a similar

III. 3β-ACETOXY-5α-CYANOCHOLESTAN-7-ONE

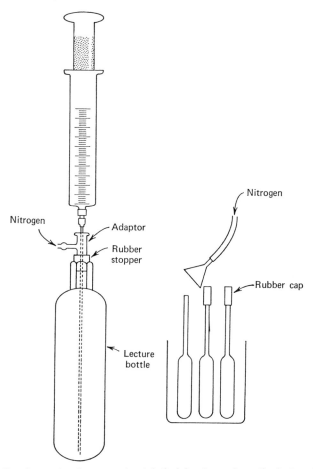

Figure 3. Apparatus for removing triethylaluminum from the lecture bottle.

procedure described by Brauer.[9] The hydrogen cyanide was collected in a tared flask and diluted with anhydrous tetrahydrofuran to a previously marked volume. The flask was capped with two serum stoppers (the second put on in an inverted position) to ensure against leakage and stored in a freezer. Solutions such as these seem to be stable for several months when kept cold.

5. Checkers obtained the steroid from K & K Laboratories.

6. A small amount of water has been found to accelerate the reaction. In the absence of water, the reaction was about 80%

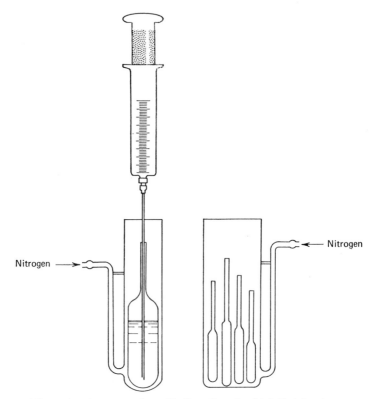

Figure 4. Apparatus for collecting aliquots of triethylaluminum.

complete at the time of 7.5 hours. Despite the rate acceleration by water, the reaction mixture should be protected from moisture, because a larger amount of water than that specified retards the reaction owing to decomposition of the triethylaluminum.

7. This treatment is an exothermic reaction with evolution of gaseous ethane. The reaction mixture should be added in a slow stream with good stirring at such a rate that the content in the flask does not overflow. When the ice has melted, additional ice should be added.

8. The reaction mixture remaining on the wall of the reaction flask is treated with a small amount of cold dilute hydrochloric acid and combined with the extraction mixture.

9. The acid treatment followed by alkaline washing of the

extracts prevents possible hydrolysis of the acetoxyl and the cyano groups.

10. The submitters have performed the extraction and washing in a countercurrent manner using three 2-l. separatory funnels.

11. The recrystallization was carried out in a 40-ml. centrifuge tube. The recrystallization mixture, after cooling to room temperature, was cooled in a freezer and washed twice with 20-ml. portions of cold recrystallization solvent.

12. When the second crop is contaminated with a polar material (α-cyanohydrin of the product) as evidenced by a lower melting point (m.p. 160–170°) and a thin-layer chromatogram (Kiesel Gel GF, benzene-ethyl acetate, 4:1, $R_f = 0.2$), the residue from the mother liquor must be treated again with alkali before crystallization.

13. This procedure is applicable to smaller or larger scale preparations with some modification. In a smaller scale experiment, the submitters suggest using a stock solution of triethylaluminum (Note 3). In a larger scale preparation, to avoid using large syringes, it is possible to run the triethylaluminum into a dry, graduated, pressure-equalizing dropping funnel. The triethylaluminum-hydrogen cyanide solution may then be just added to the reaction flask through a funnel instead of using a syringe.

14. The submitters used 5.76 g. of 3β-acetoxy-Δ⁵-cholesten-7-one and obtained a yield of 5.65 g. (92%) of the cyano ketone.

3. Discussion

3β-Hydroxy-5α-cyanocholestan-7-one has been prepared in 43% yield by the action of potassium cyanide and ammonium chloride[4] on 7-ketocholesterol at 100° for 33 hours.[5] The present method was developed by the submitters.[5]

This process is superior to classical hydrocyanation methods using an alkali metal cyanide[6] and to the improved method using potassium cyanide and ammonium chloride[4] with respect to reactivity, stereospecificity, and absence of side reactions. Also, the process is applicable to conjugate hydrocyanation of various α,β-unsaturated carbonyl and imino compounds,[5,7] cyanhydrin formation from less-reactive ketones,[5] and cleavage of epoxides to produce β-cyanohydrins,[8] the reaction conditions being varied depending on the substrate to be used. The present procedure is typical of hydrocyanation procedures using other organoaluminum compounds.

1. Shionogi Research Laboratory, Shionogi & Co., Ltd., Osaka, Japan.
2. K. Ziegler, *Org. Syn.*, Coll. Vol. **1**, 314 (1944).
3. C. W. Marshall, R. E. Ray, I. Laos, and B. Riegel, *J. Amer. Chem. Soc.*, **79**, 6308 (1957).
4. W. Nagata, S. Hirai, H. Itazaki, and K. Takeda, *J. Org. Chem.*, **26**, 2413 (1961).
5. W. Nagata, M. Yoshioka, and S. Hirai, *Tetrahedron Lett.*, 48 (1962); W. Nagata, M. Yoshioka, and M. Murakami, *J. Amer. Chem. Soc.*, **94**, 4644, 4654 (1972); W. Nagata, M. Yoshioka, and T. Terasawa, *J. Amer. Chem. Soc.*, **94**, 4672 (1972).
6. P. Kurtz, "Methoden der Organischen Chemie," Vol. 8, Georg Thieme Verlag, Stuttgart, 1952, p. 265.
7. W. Nagata, T. Okumura, and M. Yoshioka, *J. Chem. Soc. C*, 2347 (1970); W. Nagata, M. Yoshioka, T. Okumura, and M. Murakami, *J. Chem. Soc. C*, 2355 (1970).
8. W. Nagata, M. Yoshioka, and T. Okumura, *Tetrahedron Lett.*, 847 (1966); *J. Chem. Soc. C*, 2365 (1970).
9. G. Brauer, "Handbook of Preparatory Inorganic Chemistry," Vol. 1, Academic Press, New York, 1963, p. 658.

PREPARATION AND REDUCTIVE CLEAVAGE OF ENOL PHOSPHATES: 5-METHYLCOPROST-3-ENE

(5β-Cholest-3-ene, 5-methyl)

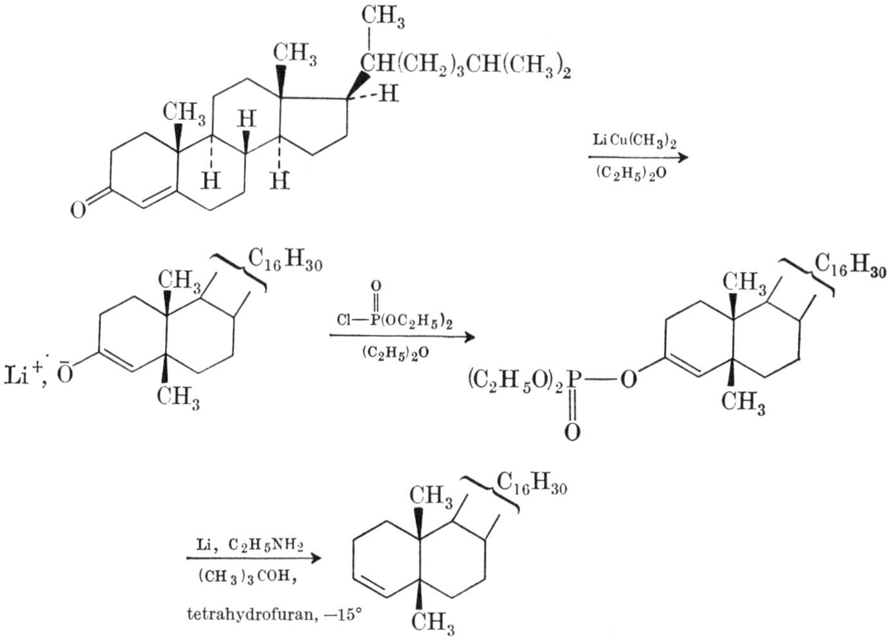

Submitted by D. C. MUCHMORE[1]
Checked by DAVID G. MELILLO and HERBERT O. HOUSE

1. Procedure

A. *Diethyl 5-Methylcoprost-3-en-3-yl Phosphate.* To a dry 100-ml. three-necked flask, equipped with a magnetic stirring bar, a pressure-equalizing dropping funnel, a nitrogen inlet tube, and a rubber septum, is added 384 mg. (0.00201 mole) of copper(I) iodide (Note 1) and 20 ml. of anhydrous ether (Note 2). After the reaction vessel has been flushed with nitrogen, a static oxygen-free nitrogen atmosphere is maintained in the reaction vessel throughout the remainder of the reaction. The reaction mixture is cooled in an ice bath and an ether solution, containing 0.0040 mole of methyllithium (Note 3), is added from a hypodermic syringe, dropwise and with stirring. As the methyllithium is added, the initial yellow precipitate of polymeric methylcopper(I) redissolves to form a colorless to pale-yellow solution of lithium dimethylcuprate (Note 4). To the resulting cold solution is added, dropwise and with stirring over 20 minutes, a solution of 576 mg. (0.00150 mole) of cholest-4-en-3-one (Note 5) in 20 ml. of ether (Note 2). During the addition of the enone, a yellow precipitate of polymeric methylcopper(I) separates from the reaction solution. After the addition is complete, the cooling bath is removed and the reaction mixture is stirred for 2 hours while it warms to room temperature. The dropping funnel in the apparatus is replaced with a second dry dropping funnel which contains a loose plug of glass wool above the stopcock. The reaction mixture is again cooled in an ice bath and a mixture of 4.0 ml. of triethylamine (Note 6) and 2.00 g. (0.0115 mole) of diethyl phosphorochloridate (Note 7) is added from the dropping funnel to the reaction mixture rapidly and with stirring. After this addition, the cooling bath is removed and stirring is continued for one hour. Saturated aqueous sodium bicarbonate solution is added to hydrolyze any remaining organometallic reagents and the reaction mixture is transferred to a separatory funnel and washed successively with two 50-ml. portions of cold (0°) aqueous $1 M$ ammonium hydroxide and with a 50-ml. portion of water. The aqueous washes are extracted in turn with a 30-ml. portion of ether and the combined ether solutions are dried over anhydrous sodium sulfate and then concentrated with a rotary evaporator. A solution of the residual liquid in 3 ml. of ether is applied to a 2.5 cm. by 15 cm. chromatographic column packed with a slurry of 50 g. of silica gel (Note 8) in ether. The column is eluted with ether. After the first 70 ml. of eluent has been collected and discarded, the next 120 ml. of ether eluent is collected and concentrated with a rotary evaporator. The residual crude phosphate ester (Note 9), a colorless liquid, amounts to 420–480 mg. and is sufficiently pure for use in the following procedure.

B. *5-Methylcoprost-3-ene.* A dry 100-ml. three-necked flask, equipped with a polyethylene-coated magnetic stirring bar, two gas inlet tubes, and a pressure-equalizing dropping funnel is immersed in a dry ice-isopropyl alcohol cooling bath maintained at $-15°$ to $-20°$. The reaction vessel is flushed with either helium or argon and a static atmosphere of one of these gases is maintained in the reaction vessel throughout the reaction. Ethylamine (Note 10) is distilled through a tower of sodium hydroxide pellets into the cold reaction flask until 50 ml. of the liquid amine has been collected. A 70-mg. (0.010 mole) piece of lithium wire is cleaned by dipping it successively into methanol and pentane and is added to the reaction flask. The resulting cold ($-15°$) mixture is stirred for 10 minutes to dissolve the lithium and then a solution of the diethyl 5-methylcoprost-3-en-3-yl phosphate and 0.50 ml. (0.39 g. or 0.0053 mole) of *t*-butyl alcohol (Note 11) in 15 ml. of tetrahydrofuran (Note 2) is added, dropwise and with stirring over 15 minutes, to the cold, blue lithium-amine solution. The resulting blue solution is stirred for an additional 15 minutes and then 1 ml. of saturated aqueous ammonium chloride solution is added to consume the excess lithium. The resulting colorless mixture is warmed to evaporate the ethylamine and the residue is diluted with 90 ml. of aqueous 10% sodium hydroxide and extracted with two 30-ml. portions of pentane. The combined organic solutions are washed with 50 ml. of aqueous sodium chloride, dried over anhydrous sodium sulfate, and then concentrated with a rotary evaporator. The residual viscous liquid is evaporatively distilled from a 25-ml. flask into a male 14–20 standard-taper glass joint as shown in Figure 1. The air bath is heated to 150–180° while the pressure in the system is maintained at 0.05 mm. to 0.4 mm. The distilled 5-methylcoprost-3-ene amounts to 260–295 mg. (45–51%) of colorless liquid, n^{25} D 1.5115–1.5123 (Note 12).

2. Notes

1. A purified grade of copper(I) iodide, purchased from Fisher Scientific Company, was used without purification.

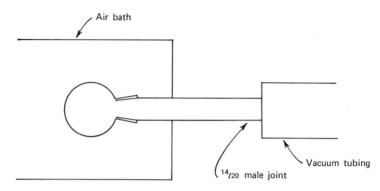

Figure 1. Apparatus for evaporative distillation.

2. Reagent grades of diethyl ether and of tetrahydrofuran were distilled from lithium aluminum hydride immediately prior to use.

3. Ethereal solutions of methyllithium are available from either Foote Mineral Company or Alpha Inorganics, Inc. These solutions should be titrated immediately before use with 2-butanol and 2,2-bipyridyl as an indicator.[2] In a typical run, 2.44 ml. of ethereal 1.64M methyllithium was employed.

4. The appearance of a brown to black precipitate indicates either oxidative or thermal decomposition of the cuprate. If such decomposition has occurred, it is best to prepare the reagent again with greater care to avoid molecular oxygen and/or excessive reaction temperatures.

5. A commercial sample of cholest-4-en-3-one from Eastman Organic Chemicals was used without further purification. The preparation of this ketone has also been described in *Organic Syntheses*.[3]

6. The triethylamine is used to trap any hydrogen chloride present as the insoluble triethylammonium chloride which is collected in the glass wool filter. A reagent grade of triethylamine (b.p. 88°) was distilled from calcium hydride prior to use.

7. Commercial diethyl phosphorochloridate [b.p. 60–62° (1.5 mm.)], purchased from Eastman Organic Chemicals, was distilled prior to use in this reaction.

8. A good grade of silica gel, such as that available from E.

Merck and Company, Darmstadt, is appropriate for this chromatography.

9. The n.m.r. spectrum ($CDCl_3$ solution) of the crude product has absorption at 5.1 (1H multiplet, vinyl CH), 3.9–4.4 (4H multiplet, CH_2—O), and 0.6–2.3 p.p.m. (ca. 52H multiplet, aliphatic CH).

10. Ethylamine (b.p. 17°) is available from Eastman Organic Chemicals.

11. A commercial grade of t-butyl alcohol (b.p. 83°) should be distilled from calcium hydride before use.

12. The product exhibits end absorption in the ultraviolet (95% EtOH solution) with ϵ 330 at 210 mμ and a series of n.m.r. peaks ($CDCl_3$ solution) at 0.67, 0.82, 0.85, 0.88, and 0.92 p.p.m. (18H, CH_3) with a multiplet at 1.0–2.2 p.p.m. and a partially resolved multiplet attributable to two vinyl protons. This latter absorption corresponds approximately to signals at 5.28 (1H doublet of triplets, $J = 8$ and 1 Hz.) and 5.60 p.p.m. (1H doublet of triplets, $J = 8$ and 2.8 Hz.). The mass spectrum of the product has the following abundant peaks: m/e (rel. int.), 384 (100, M+), 369 (69), 355 (70), 229 (27), 122 (28), 109 (28), 107 (60), 95 (33), 93 (34), 81 (72), 55 (30), and 43 (31).

The submitters report that ozonolysis of the product at −10° in a mixture of ethyl acetate and acetic acid, followed by reaction with hydrogen peroxide, formed 3,4-seco-5-methylcoprostan-3,4-dioic acid as crystals from ethyl acetate, m.p. 168–172° with prior softening at 130°.

3. Discussion

The conjugate addition of lithium dimethylcuprate and other organocopper reagents to α,β-unsaturated ketones is a reaction which has had wide application and which has been fairly well studied.[5] In order that the positional specificity which has been conferred upon the enolate anions generated by such additions might be maintained, these intermediates have been intercepted with acetic anhydride,[5a] chlorotrimethylsilane,[6] and diethyl phosphorochloridate.[4,9]

The reductive fission of enol phosphates to form olefins is a modification of the procedure used by Kenner and Williams[7] to deoxygenate phenols. The enol phosphates, which have been reduced by the action of lithium in ammonia or alkylamines, have been prepared by treatment of α-bromoketones with triethyl phosphite,[8,9] by interception of enolates generated by the addition of lithium dimethylcuprate to α,β-unsaturated ketones,[4,9] and by interception of enolates resulting from treatment of unsaturated ketones with lithium in ammonia.[4]

1. Division of Chemistry and Chemical Engineering, Gates & Crellin Laboratories, California Institute of Technology, Pasadena, California 91109.
2. (a) S. C. Watson and J. F. Eastham, *J. Organometal. Chem.*, **9**, 165 (1967); (b) W. Voskuil and J. F. Arens, *Org. Syn.*, **48**, 47 (1968); (c) M. Gall and H. O. House, *Org. Syn.*, **52**, 40 (1972).
3. J. F. Eastham and R. Teranishi, *Org. Syn.*, Coll. Vol. **4**, 192 (1963).
4. R. E. Ireland and G. Pfister, *Tetrahedron Lett.*, 2145 (1969).
5. (a) H. O. House, W. L. Respess, and G. M. Whitesides, *J. Org. Chem.*, **31**, 3128 (1966); (b) H. O. House and W. F. Fischer, Jr., *J. Org. Chem.*, **33**, 949 (1968); (c) G. Posner, *Org. React.*, **19**, 1 (1972).
6. (a) G. Stork and P. F. Hudrlik, *J. Amer. Chem. Soc.*, **90**, 4462 (1968); (b) H. O. House, L. J. Czuba, M. Gall, and H. D. Olmstead, *J. Org. Chem.*, **34**, 2324 (1969).
7. G. W. Kenner and N. R. Williams, *J. Chem. Soc.*, 522 (1955).
8. M. Fetizon, M. Jurion, and N. T. Anh, *Chem. Commun.* 112 (1969).
9. D. Muchmore, Ph.D. Thesis, California Institute of Technology, 1971.

REACTION OF ARYL HALIDES WITH π-ALLYLNICKEL HALIDES: METHALLYLBENZENE

(Benzene, 2-methylalkyl)

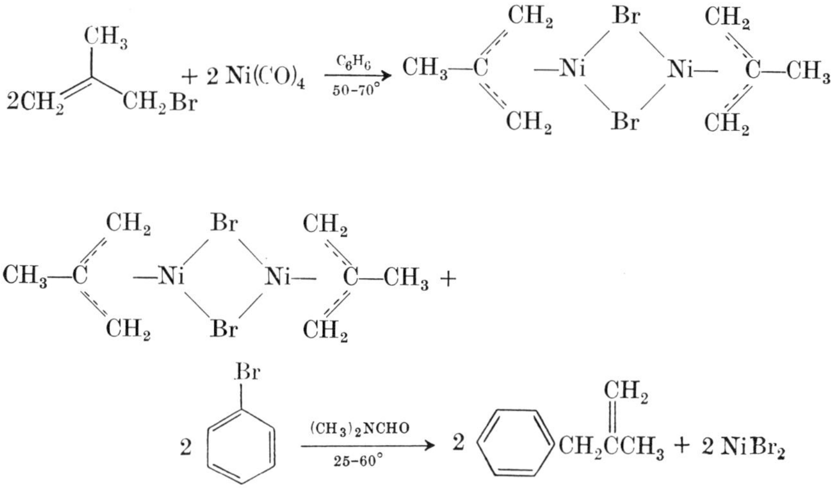

$$2CH_2\!\!=\!\!\underset{CH_2Br}{\overset{CH_3}{C}} + 2\,Ni(CO)_4 \xrightarrow[50-70°]{C_6H_6} CH_3\!-\!\underset{CH_2}{\overset{CH_2}{C}}\!\!\diagdown\!\underset{Br}{\overset{}{Ni}}\!\diagdown\!\underset{Br}{\overset{}{Ni}}\!-\!\underset{CH_2}{\overset{CH_2}{C}}\!-\!CH_3$$

$$CH_3\!-\!\underset{CH_2}{\overset{CH_2}{C}}\!\!\diagdown\!\underset{Br}{\overset{}{Ni}}\!\diagdown\!\underset{Br}{\overset{}{Ni}}\!-\!\underset{CH_2}{\overset{CH_2}{C}}\!-\!CH_3\ +\ 2\,\underset{}{\overset{Br}{\bigcirc}} \xrightarrow[25-60°]{(CH_3)_2NCHO} 2\,\bigcirc\!\!-\!CH_2\overset{CH_2}{\underset{\|}{C}}CH_3 + 2\,NiBr_2$$

Submitted by MARTIN F. SEMMELHACK and PAUL M. HELQUIST[1]
Checked by BRADLEY E. MORRIS and RICHARD E. BENSON

1. Procedure

Caution! Nickel carbonyl is a flammable, volatile (b.p. 43°), highly toxic reagent. Safety glasses, gloves, and an apron should be worn when handling this reagent and the first step of this preparation should be conducted in an efficient hood (Note 1).

A 1-l. three-necked flask is equipped with a reflux condenser, a pressure-equalizing dropping funnel, a three-way stopcock, and a large magnetic stirring bar. The system is flushed with argon (Note 2) and 380 ml. of benzene (Note 3) is placed in the flask. From an inverted lecture cylinder 50.8 g. (38.5 ml., 0.298 mole) of nickel carbonyl (Note 4) is introduced into the addition funnel. The nickel carbonyl is then added to the benzene while an atmosphere of argon is maintained, and the flask is immersed in an oil bath at 50°. By means of a syringe, 10.04 g. (0.0745 mole) of methallyl bromide (Note 5) is added over 10 minutes. After a short induction period, evolution of carbon monoxide becomes rapid and a deep red color appears. The exit gas is led from the top of the condenser through a gas bubbler tube in order to monitor the rate of gas evolution. As the gas evolution becomes vigorous, the bath temperature is raised to 70° and maintained at this temperature for 30 minutes after gas evolution ceases (the total time after addition of methallyl bromide is 1.5 hours). The resulting solution is allowed to cool to 25°, and the benzene and excess nickel carbonyl are removed under reduced pressure (water aspirator) applying an oil bath at 30° as needed to maintain a rapid rate of evaporation (Note 6). The red solid residue ($>85\%$ yield) is π-methallylnickel bromide which is used directly in the next step (Note 7).

To a solution of the crude nickel complex (an 85% yield is assumed) in 65 ml. of oxygen-free N,N-dimethylformamide (Note 8) under an argon atmosphere at 25° is added a solution of 9.95 g. (0.0634 mole) of bromobenzene (Note 9) in 100 ml. of oxygen-free N,N-dimethylformamide (Note 8) during a 15-minute period. After the addition is complete, the reaction mixture is stirred at 25° for 12 hours, followed by warming to 60° for one hour. Complete reaction of the nickel complex is indicated by a color change from red to the emerald green color characteristic of a solution of nickel dibromide in dimethylformamide. After being cooled to 25°, the solution is poured into a mixture of 250 ml. of water and 250 ml. of petroleum ether (b.p. 30–60°); 2 ml. of aqueous $12M$ hydrochloric acid is added (Note 10) and the mixture is filtered through Celite filter aid to facilitate separation of the layers. The organic layer is separated, washed with two 100-ml. portions of water, dried over anhydrous magnesium sulfate, and concentrated by use of a rotary evaporator at water aspirator pressure to afford 8.0–9.6 g. of a clear, colorless liquid. Distillation through a short Vigreux column gives 5.58–6.02 g. (67–72% yield based on bromobenzene) of methallylbenzene as a colorless liquid, b.p. 67–68° (19 mm.), n^{25} D 1.5064 (Note 11).

2. Notes

1. The treatment for nickel carbonyl poisoning involves intramuscular injection of BAL (2,3-dimercapto-1-propanol).[2]

2. Argon is preferred by the submitters for its high density which allows opening of the reaction vessel without significant displacement of the inert atmosphere by air. A nitrogen atmosphere would be equally effective in preventing oxidation of the π-allylnickel complex. The checkers used a nitrogen atmosphere.

3. Anhydrous, air-free benzene was prepared by distillation under argon, discarding a 20% forerun. The checkers used benzene from a freshly opened bottle (Fisher Scientific Company).

4. Nickel carbonyl available from Matheson Gas Products was used by the checkers.

5. Methallyl bromide is prepared from methallyl chloride (Eastman Organic Chemicals) by means of a halide exchange reaction. A solution of 148.3 g. (1.64 moles) of methallyl chloride and 213.8 g. (2.48 moles) of lithium bromide in 1 l. of dry acetone is refluxed for 5 hours. The mixture is filtered and the filtrate is distilled through a 30-cm. Vigreux column to afford 69.2 g. (31.3%) of methallyl bromide, b.p. 88–93°, n^{25} D 1.4672. The purity was 98 ± 2% by gas chromatographic analysis on a 20% Carbowax column at 65°.

6. Nickel carbonyl is drawn into the aspirator flow during this operation. In many laboratories the hood plumbing is connected with the general plumbing line and vapors of highly toxic nickel carbonyl may diffuse back to sinks at the laboratory bench. If such an arrangement is suspected, the solvent and excess nickel carbonyl can be collected by employing a cold trap ($-78°$ or $-196°$) between the reaction mixture and the aspirator. Care should be used in the disposal of this mixture.

7. Pure π-methallylnickel bromide can be obtained by dissolving the residue in 150 ml. of oxygen-free anhydrous ether, filtering under argon, concentrating the filtrate until crystals begin to form, and cooling to $-78°$ for 12 hours. The crystals are

isolated by removing the liquid via suction through a syringe needle under a positive pressure of argon. The yield is 12.1 g. (85%) of dark red crystals. The n.m.r. spectrum can be obtained only by rigorous exclusion of oxygen from the sample and filtration of the sample as the last stage of sample preparation. The n.m.r. spectrum, determined in benzene solution, shows three singlets at 2.83 (2H), 2.82 (2H), and 2.07 p.p.m. (3H).

8. N,N-Dimethylformamide is distilled from calcium hydride at 71° (32 mm.) and stored under argon. The checkers used a freshly opened bottle of the product (white label grade) available from Eastman Organic Chemicals.

9. Bromobenzene was used as supplied by Aldrich Chemical Company, Inc. Purification by distillation under argon did not change the yield of methallylbenzene. The checkers used the product available from Eastman Organic Chemicals (white label).

10. The hydrochloric acid solution is added to speed solution of the nickel salts that otherwise lead to emulsions during separation. If no emulsion is encountered after mixing the petroleum ether and water solutions, no hydrochloric acid need be added. Similarly, the filtration through Celite filter aid is intended to remove finely divided nickel metal and other insoluble particles which complicate the washing procedure. If no particles are present, the filtration step should be omitted.

11. The product consists of 99% methallylbenzene and 1% 2,5-dimethyl-1,5-hexadiene by n.m.r. spectroscopic analysis. The n.m.r. spectrum of methallylbenzene (CCl_4 solution) shows peaks at 7.15 (singlet, 5H, phenyl), 4.75 (multiplet, 2H, vinyl CH), 3.25 (broad singlet, 2H, allylic CH_2), and 1.63 p.p.m. (broad singlet, 3H, allylic CH_3). The n.m.r. spectrum of 2,5-dimethyl-1,5-hexadiene (CCl_4 solution) has peaks at 4.75 (multiplet, 4H, vinyl), 2.12 (broad singlet, 4H, allylic CH_2), and 1.70 p.p.m. (broad singlet, 6H, allylic CH_3). A small forerun contained 0.30 g. (3% yield) of methallylbenzene and a larger quantity of 2,5-dimethyl-1,5-hexadiene. The distillation residue is composed of 0.24–0.34 g. (3–4% yield) of methallylbenzene and 0.38–0.52 g. (8–10% yield) of biphenyl. The distillation

fractions may be analyzed by gas chromatography by use of a 180 cm. by 6.4 mm. column packed with 10% SE-30 on Chromosorb G. The retention times for methallylbenzene and 2,5-dimethyl-1,5-hexadiene are 4.0 minutes and 2.0 minutes, respectively, at 125°.

3. Discussion

The simple example outlined above of replacement of halogen by a methallyl group could be carried out in an equally direct way using phenylmagnesium bromide and methallyl halide. However, the Grignard reaction is complicated by formation of the conjugated isomer β,β-dimethylstyrene,[3] or by a rearrangement to *trans*-2-butenylbenzene.[4] In no case has this approach afforded methallylbenzene in greater than 50% yield. Dehydration of (2-hydroxy-2-methylpropyl)benzene also produces a mixture of methallylbenzene (68%) and β,β-dimethylstyrene (32%).[5] Elimination of benzoic acid from the benzoate ester of (2-hydroxy-2-methylpropyl)benzene gives the same ratio of products, although the combined yield is lower, (86%).[6] The Wittig reaction of methylenetriphenylphosphorane with 1-phenyl-2-propanone produces methallylbenzene in only 2% yield.[7]

The preparation illustrates the procedure for formation of π-allylnickel halides and their reaction with aryl halides.[8] The complexes can be obtained from allylic chlorides, bromides, and iodides[9–11] even when the allylic halides bears substituents such as alkyl, carboalkoxyl, and alkenyl side chains.[10] The coupling step is generally applicable to aryl, alkyl, and vinyl bromides or iodides;[8] organic chlorides are usually unreactive with π-allylnickel halides. Other polar aprotic solvents (hexamethylphosphoramide, dimethyl sulfoxide, N-methylpyrrolidone) have been used.[12] Protic solvents lead to destruction of the π-allylnickel complex by slow protonation of the allyl ligand.[12] No reaction occurs between aryl, alkyl, or vinyl halides and π-allylnickel bromide in less polar solvents such as tetrahydrofuran, 1,2-dimethoxyethane, ether, or hydrocarbons. The π-allylnickel bromide complexes are very reactive with allyl halides but halogen-metal exchange precedes coupling

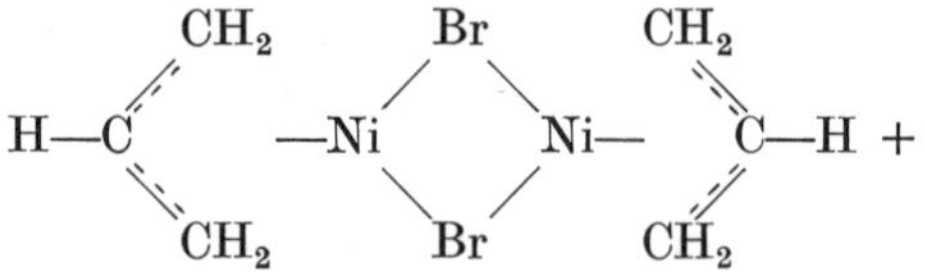

π-allylnickel bromide

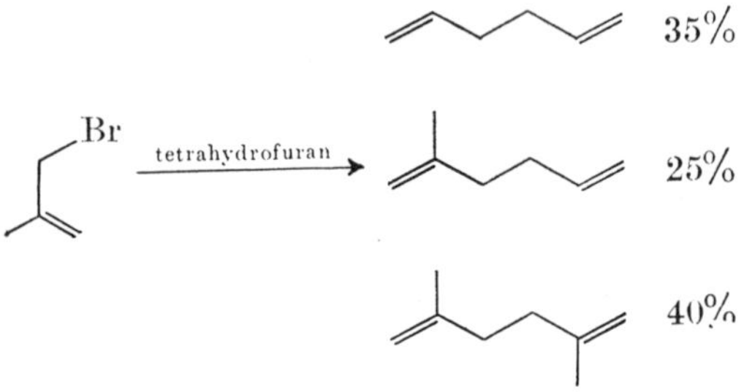

and a mixture of products is obtained as illustrated in the accompanying example.[13] The π-allylnickel complex from *trans*-geranyl bromide reacts with alkyl halides to give a mixture of *cis* and *trans* products;[8] the double bond that participates in the π-allyl group is isomerized during the sequence of reactions:

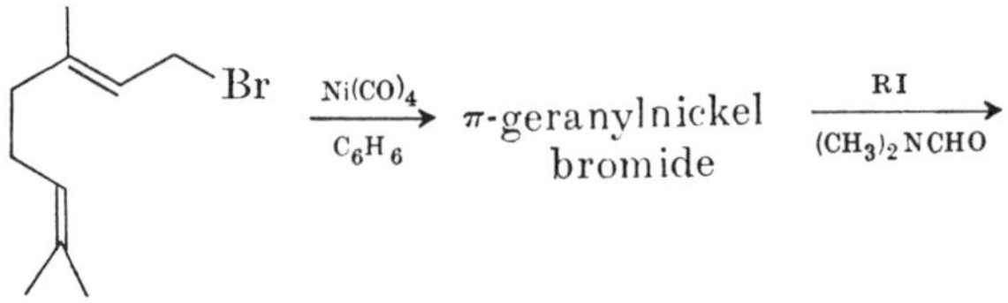

trans-geranyl bromide

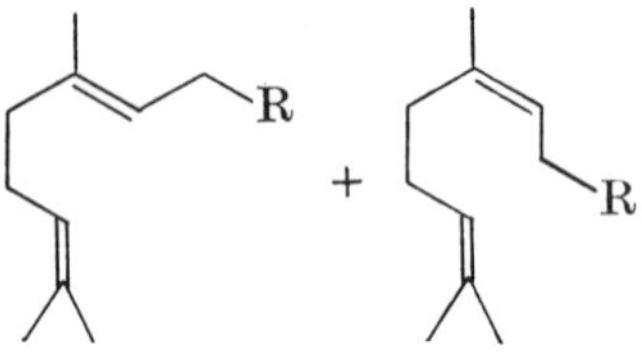

trans 43% *cis* 35%

R = cyclohexyl

Similarly, *trans*-4-iodocyclohexanol reacts with π-methallylnickel bromide to produce a mixture of epimeric 4-methallylcyclohexanols.[8] It is important to note that the hydroxyl group has no significant effect on this reaction.

The advantages of the π-allylnickel halides reside in their nonnucleophilic and nonbasic character which allow especially selective reactions with organic halides in the presence of a large number of other functional groups. Carbonyl and hydroxyl groups react much more slowly than organic halides (especially iodides) with π-allylnickel halides while olefins, nitriles, alkyl chlorides, and aromatic hydrocarbons are inert to these reagents.[8,12]

1. Department of Chemistry, Cornell University, Ithaca, New York 14850.
2. P. G. Stecher, "The Merck Index," 8th ed., Merck and Co., Inc., Rahway, N.J., 1968, p. 372.
3. C. M. Buess, J. V. Karavinos, P. V. Kunz, and L. C. Gibbons, *Nat. Adv. Comm. Aeronaut. Tech. Notes*, No. 1021 (1946).
4. K. W. Wilson, J. D. Roberts, and W. G. Young, *J. Amer. Chem. Soc.*, **71**, 2019 (1949).
5. L. Beránek, M. Kraus, K. Kochloefl, and V. Bažant, *Collect. Czech. Chem. Commun.*, **25**, 2513 (1960) [*C.A.*, **55**, 3464 (1961)].
6. R. Onesta and G. Castelfranchi, *Chim. Ind. (Milan)*, **42**, 735 (1960) [*C.A.*, **55**, 24646 (1961)].
7. C. Rüchardt, *Chem. Ber.*, **94**, 2599 (1961).
8. E. J. Corey and M. F. Semmelhack, *J. Amer. Chem. Soc.*, **89**, 2755 (1967).
9. E. O. Fischer and G. Bürger, *Z. Naturforsch. B*, **16**, 77 (1961).
10. M. F. Semmelhack, Ph.D. Dissertation, Harvard University, 1967.
11. G. Wilke, B. Bogdanović, P. Hardt, P. Heimbach, W. Keim, M. Kröner, W. Oberkirch, K. Tanaka, E. Steinrücke, D. Walter, and H. Zimmermann, *Angew. Chem., Int. Ed. Engl.*, **5**, 151 (1966).
12. See reference 10 and unpublished results of M. F. Semmelhack and E. J. Corey.
13. E. J. Corey, M. F. Semmelhack, and L. S. Hegedus, *J. Amer. Chem. Soc.*, **90**, 2416 (1968).

REDUCTION OF KETONES BY USE OF THE TOSYLHYDRAZONE DERIVATIVES: ANDROSTAN-17-β-OL

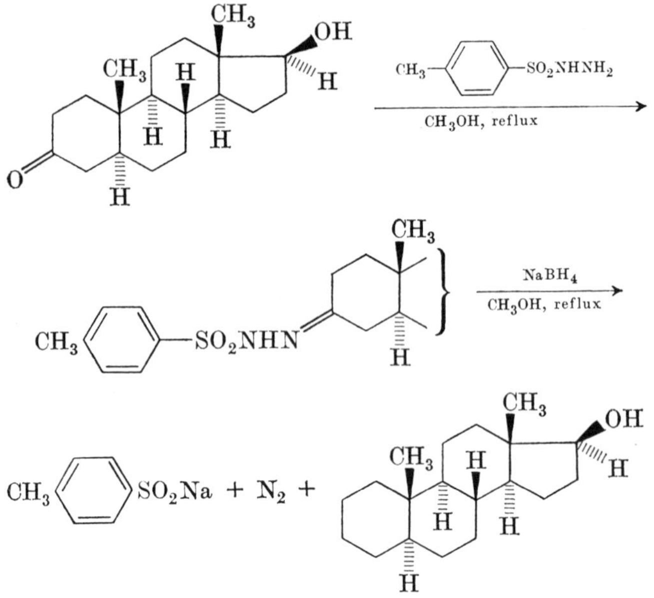

Submitted by L. CAGLIOTI[1]
Checked by J. F. MOSER and A. ESCHENMOSER

1. Procedure

In a 100-ml. round-bottomed flask equipped with a reflux condenser are placed 1.00 g. (0.0035 mole) of 5α-androstan-17β-ol-3-one (Note 1), 0.90 g. (0.0048 mole) of tosylhydrazide (Note 2), and 70 ml. of methanol (Note 3). The mixture is heated under gentle reflux for 3 hours then cooled to room temperature. To the solution is added 2.5 g. (0.075 mole) of sodium borohydride in small portions during one hour (Note 4) and the resulting mixture is heated under reflux for an additional 8 hours. The reaction mixture is cooled to room temperature and the solvent is removed under reduced pressure. The residue is dissolved in ether, transferred to a separatory funnel, and washed successively with water, dilute aqueous sodium carbonate, aqueous 2M hydrochloric acid, and water. The ethereal solution is dried over anhydrous sodium sulfate and the solvent is evaporated under reduced pressure. The residue, 0.95 g. of white crystals (Note 5), is dissolved in about 20 ml. of a 7:3 (v/v) mixture of cyclohexane and ethyl acetate and the resulting solution is applied to a column prepared from 60 g. of silica gel (Merck, 0.05–0.2 mm.). The column is eluted with the 7:3 (v/v) cyclohexane-ethyl acetate mixture and a 200-ml. fraction is collected. Evaporation of this fraction under reduced pressure affords 0.70–0.73 g. (73–76%) of pure 5α-androstan-17 β-ol. Crystallization from aqueous methanol provides 0.64 g. of analytically pure product, m.p. 161–163°.

2. Notes

1. 5α-Androstan-17β-ol-3-one was supplied by Aldrich Chemical Co., Inc.

2. Tosylhydrazide was supplied by Aldrich Chemical Co., Inc. Alternately, it may be prepared by a procedure described in *Organic Syntheses*.[6]

3. Tetrahydrofuran serves equally well as a solvent. However the quantity of sodium borohydride should be reduced to 1.0 g. and the isolation procedure modified in the following way. After the solution has been refluxed for 8 hours, the reaction mixture is cooled and the excess sodium borohydride is decomposed by the slow addition of dilute aqueous hydrochloric acid. The resulting mixture is extracted with ether and the ethereal solution is washed as described.

4. Because of the ready decomposition of sodium borohydride in methanol, the solution is maintained at room temperature during the addition of the metal hydride.

5. The crude product is contaminated with a small amount of a more polar substance which is subsequently removed by chromatography.

3. Discussion

The preparation of 5α-androstan-17β-ol from 5α-androstan-17β-ol-3-one may be realized by classical methods such as the Wolff-Kishner or Clemmensen reduction.

This procedure illustrates a general method for the reduction of aldehyde and ketone functions to methylene groups under very mild conditions. Since strong acids and bases are not employed, this procedure is of particular importance for the reduction of ketones possessing an adjacent chiral center.[2,3] Moreover, the use of deuterated metal hydrides permits the preparation of labelled compounds.[4]

The reduction of the preformed tosylhydrazones with sodium borohydride may be effected in aprotic solvents, such as tetrahydrofuran or dioxane.[5] The use of lithium aluminium hydride in nonhydroxylic solvents permits the reduction of aromatic aldehydes and ketones.

1. Istituto di Chimica Organica, Universita di Roma, Rome, Italy.
2. The reduction of the tosylhydrazone of (+)(S)-4-methyl-3-hexanone affords (+)(S)-3-methylhexane, optical purity 85%; L. Lardicci and C. Botteghi, private communication.
3. A. N. De Belder and R. Weigel, *Chem. Ind.* (*London*), 1689 (1964).
4. E. J. Corey and S. K. Gros, *J. Amer. Chem. Soc.*, **89**, 4561 (1967); M. Fischer, Z. Pelah, D. H. Williams, and C. Djerassi, *Chem. Ber.*, **98**, 3236 (1965).
5. L. Caglioti, *Tetrahedron*, **22**, 487 (1966).
6. L. Friedman, R. L. Litle, and W. R. Reichle, *Org. Syn.*, **40**, 93 (1960).

REDUCTIVE AMINATION WITH SODIUM CYANOBOROHYDRIDE: N,N-DIMETHYLCYCLOHEXYL-AMINE

$$\text{cyclohexanone} + (CH_3)_2NH \cdot HCl + KOH \xrightarrow[CH_3OH]{NaBH_3CN} \text{N,N-dimethylcyclohexylamine}$$

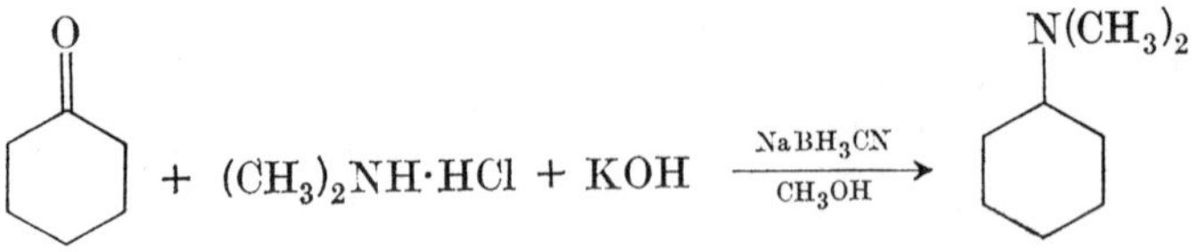

Submitted by RICHARD F. BORCH[1]
Checked by K. ABE and S. MASAMUNE

1. Procedure

A solution of 21.4 g. (0.25 mole) of dimethylamine hydrochloride in 150 ml. of methanol is prepared in a 500-ml. round-bottomed flask. Potassium hydroxide (4 g.) is added in one portion to the magnetically stirred solution (Note 1). When the pellets are completely dissolved, 19.6 g. (0.20 mole) of cyclohexanone is added in one portion. The resulting suspension is stirred at room temperature for 15 minutes, and then a solution of 4.75 g. (0.076 mole) of sodium cyanoborohydride (Notes 2 and 3) in 50 ml. of methanol is added dropwise over 30 minutes to the stirred suspension. After the addition is complete, the suspension is stirred for 30 minutes. Potassium hydroxide (15 g.) is then added, and stirring is continued until the pellets are completely dissolved. The reaction mixture is filtered with suction and the volume of the filtrate is reduced to approximately 50 ml. with a rotary evaporator while the bath temperature is kept below 45° (Notes 4 and 5). To this concentrate is added 10 ml. of water and 25 ml. of saturated aqueous sodium chloride and the layers are separated. The aqueous layer is extracted with two 50-ml. portions of ether. The organic layer previously separated and the ethereal extracts are combined and extracted with three 20-ml. portions of aqueous $6M$ hydrochloric acid (Note 6). The combined acid layers are saturated with sodium chloride and extracted with four 30-ml. portions of ether (Note 7). The aqueous solution is cooled to 0° in an ice bath and brought to pH >12 by addition of potassium hydroxide pellets to the stirred solution (Notes 8 and 9). The layers are separated, and the aqueous layer is extracted with two 40-ml. portions of ether. The combined organic layers are washed with 10 ml. of saturated aqueous sodium chloride solution, dried over anhydrous potassium carbonate, and freed of ether with a rotary evaporator (Note 4). This crude product is fractionated through a 15-cm. Vigreux column (Note 10). After 1–3 g. of a forerun, b.p. 144–155° (Note 11) is separated, the fraction boiling at 156–159° is collected to give 13.3–13.7 g. (52–54%) of N,N-dimethylcyclohexylamine, n^{25} D 1.4521 (Note 12).

2. Notes

1. Precipitation of potassium chloride begins immediately; the presence of this solid does not interfere with the reaction, and removal by filtration will result in loss of dimethylamine.

2. Sodium cyanoborohydride is available as a pale brown solid from Alfa Inorganics, Inc.

3. The commercially available material can be used without further purification. Use of material purified by the published procedure[2] gives a less colored crude product, but makes no improvement in yield or purity of the final product.

4. Since the product boils at 75° (15 mm.), care should be exercised to prevent loss of material in the evaporation process.

5. It is normal for additional potassium chloride to precipitate as the evaporation continues.

6. *Caution! This addition of hydrochloric acid into a separatory funnel occurs with considerable heat evolution, causing the ether to boil. The initial addition must be carried out with gentle swirling and cooling.*

7. Gas chromatographic analysis shows that the ethereal extract contains solely cyclohexanol ($>98\%$).

8. The aqueous layer in this step is saturated with ether, and the addition of potassium hydroxide must be carried out gradually to prevent the contents of the flask from boiling over.

9. Copious amounts of potassium chloride precipitate during this addition. It is not necessary to remove the salt by filtration before the ether extraction.

10. A still pot of at least 100-ml. capacity should be used for the distillation since foaming occurs as the distillation proceeds.

11. On a 2-m. gas chromatography column packed with 10% Apiezon L and heated to 100°, the retention times for N,N-dimethylcyclohexylamine and cyclohexanol are 15 and 4 minutes, respectively. The composition of this forerun is 80–85% of the amine and 20–15% of the alcohol.

12. Gas chromatographic analysis of the product shows that the product is at least 99.2% pure and is contaminated only with trace amounts of cyclohexanol. The submitter reported a 62–69% yield (15.7–17.5 g.) using the indicated scale.

3. Discussion

N,N-Dimethylcyclohexylamine has been prepared by catalytic reductive alkylation[3,4] and by the Leuckart reaction.[5]

TABLE I
REPRESENTATIVE REDUCTIVE AMINATIONS WITH $NaBH_3CN$[2]

Compound	Amine	Product	Yield, %
Cyclohexanone	NH_3	Cyclohexylamine	45
Cyclohexanone	CH_3NH_2	N-Methylcyclohexylamine	41
Cyclohexanone	CH_3NHCH_3	N,N-Dimethylcyclohexylamine	53
Acetophenone	NH_3	α-Phenylethylamine	77
Acetophenone	CH_3NH_2	N-Methylphenethylamine	78
Isobutyraldehyde	$PhNH_2$	N-Isobutylaniline	78
Glutaraldehyde	CH_3NH_2	N-Methylpiperidine	43

The present method is experimentally simple, requires no special apparatus, and is generally applicable to the synthesis of a variety of primary, secondary, and tertiary amines as illustrated in Table I.

The submitter has found that use of $NaBH_4$ instead of $NaBH_3CN$ in the present procedure results in the almost exclusive formation of cyclohexanol with less than 3% of basic material.

1. Department of Chemistry, University of Minnesota, Minneapolis, Minnesota 55455.
2. R. F. Borch, M. D. Bernstein, and H. D. Durst, *J. Amer. Chem. Soc.*, **93**, 2897 (1971).
3. J. D. Roberts and V. C. Chambers, *J. Amer. Chem. Soc.*, **73**, 5030 (1951).
4. W. S. Emerson, *Org. React.*, **4**, 174 (1948).
5. R. D. Bach, *J. Org. Chem.*, **33**, 1647 (1968).

SUBSTITUTION OF ARYL HALIDES WITH COPPER(I) ACETYLIDES: 2-PHENYLFURO[3,2-*b*]PYRIDINE

$$Cu^{II}(NH_3)_4{}^{+2} \xrightarrow{HONH_3{}^+ \; Cl^-} Cu^I(NH_3)_2{}^+$$

$$Cu^I(NH_3)_2{}^+ + C_6H_5\!-\!C\!\equiv\!CH \longrightarrow C_6H_5\!-\!C\!\equiv\!C\!-\!Cu + NH_4{}^+ + NH^3$$

[2-iodopyridin-3-ol] $+ \; C_6H_5\!-\!C\!\equiv\!C\!-\!Cu \xrightarrow[\text{reflux}]{\text{pyridine}}$ [2-phenylfuro[3,2-*b*]pyridine] $-C_6H_5 + CuI$

Submitted by D. C. OWSLEY and C. E. CASTRO[1]
Checked by MICHAEL J. UMEN and HERBERT O. HOUSE

1. Procedure

A. *Copper(I) Phenylacetylide.* In a 2-l. Erlenmeyer flask, fitted with a large magnetic stirring bar (Note 1) and an ice-water cooling bath, is placed a solution of 25.0 g. (0.100 mole) of copper(II) sulfate pentahydrate (Note 2) in 100 ml. of concentrated aqueous ammonia. The solution is stirred with cooling for 5 minutes while a stream of nitrogen gas is passed over the solution (Note 3). Then 400 ml. of water is added and stirring and cooling under a nitrogen atmosphere (Note 3) are continued for 5 minutes. Solid hydroxylamine hydrochloride (13.9 g. or 0.200 mole, Note 4) is added to the reaction solution, with continuous stirring and cooling under nitrogen, over 10 minutes (Note 5) and then a solution of 10.25 g. (0.100 mole) of phenylacetylene (Note 6) in 500 ml. of 95% ethanol is added rapidly to the pale blue solution. The reaction flask is swirled by hand during the separation of the copper(I) phenylacetylide as a copious yellow precipitate and then an additional 500 ml. of water is added. After the mixture has been allowed to stand for 5 minutes, the precipitate is collected on a sintered glass filter (Note 7) and then washed successively with five 100-ml. portions of water, with five 100-ml. portions of absolute ethanol, and with five 100-ml. portions of anhydrous ether. The copper(I) acetylide is dried by placing it in a 250-ml. round-bottom flask which is heated to 65° for 4 hours under reduced pressure on a rotary evaporator. The yield of the copper(I) phenylacetylide, a bright yellow solid, is 14.8–16.4 g. (90–99%). The dry acetylide may be stored under nitrogen in a brown bottle (Note 8).

B. *2-Phenylfuro [3,2-b] Pyridine.* To a 300-ml. three-necked flask, fitted with a nitrogen inlet stopcock, a magnetic stirring bar, and a condenser attached to a nitrogen outlet stopcock and a mercury trap, is added 2.47 g. (0.0150 mole) of copper(I) phenylacetylide. The system is purged with nitrogen for 20 minutes and then 80 ml. of pyridine (Note 9) is added. The resulting mixture is stirred for 20 minutes under a nitrogen atmosphere (Note 10) and then 3.30 g. (0.0149 mole) of 2-iodo-3-pyridinol (Note 11) is added. The mixture, which changes in color from yellow to dark green as the acetylide dissolves (Note 12), is warmed in an oil bath at 110–120° for 9 hours with continuous stirring under a nitrogen atmosphere (Note 10). The reaction solution is then transferred to a 500-ml. round-bottom flask and concentrated to a volume of 20 ml. at 60–70° and 20–80 mm. pressure with a rotary evaporator. The pyridine solution is treated with 100 ml. of concentrated aqueous ammonia and the resulting deep blue mixture is stirred for 10 minutes and then extracted with five 100-ml. portions of ether. The combined ethereal extracts are washed with three 250-ml. portions of water and then dried over anhydrous magnesium sulfate and concentrated with a rotary evaporator. The crude product, 2.6–2.76 g. of orange semisolid, is dissolved in 100 ml. of boiling cyclohexane, and the solution is filtered, concentrated to a volume of about 30 ml., and cooled in an ice bath. The partially purified product crystallizes as 2.3–2.7 g. of orange solid, m.p. 83–89°. Further purification is effected by sublimation at 110–120° and 0.01–0.2 mm. The product, a yellow solid melting at 90–91°, amounts to 2.2–2.4 g. (75–82%) (Note 13).

2. Notes

1. An 8-cm. Teflon-coated stirring bar is convenient.

2. A reagent grade copper(II) sulfate pentahydrate, purchased from either Mallinckrodt Chemical Works or J. T. Baker Chemical Company, may be employed.

3. A nitrogen atmosphere is maintained above the reaction

solution throughout the preparation of the copper(I) acetylide.

4. Material of satisfactory purity was obtained either from J. T. Baker Chemical Company or from Coleman Matheson and Bell.

5. Too rapid an addition of the hydroxylamine salt results in precipitation of a dark solid that dissolves slowly. If solids do separate, they should be pulverized to hasten solution.

6. Phenylacetylene, purchased either from K & K Laboratories or from Aldrich Chemical Company, Inc., was used without purification.

7. A 600-ml. coarse porosity sintered glass filter is recommended to shorten the filtration time. The filtration may also be hastened by periodically scraping the bottom of the funnel with a spatula.

8. The submitters report that the acetylide is stable for years under these conditions.

9. A reagent grade of pyridine, purchased from either J. T. Baker Chemical Company or Fisher Scientific Company, was employed.

10. Oxygen will convert the acetylide to 1,4-diphenylbutadiyne.[4a]

11. This material, obtained from Aldrich Chemical Company, Inc., was used without purification.

12. Although the reaction mixture becomes homogeneous in this example, the submitters report that only partial solution occurs in other successful substitution reactions. The solubilities of the acetylides and the heterogeneous character of the cyclization have been described.[7]

13. The product exhibits ultraviolet maxima (95% EtOH solution) at 312 mμ (ϵ 32,900) and 326 mμ (ϵ 27,100) with n.m.r. peaks (acetone-d_6 solution) at 8.49 (1H, doublet of doublets, $J = 1.4$ and 4.7 Hz.) and 7.1–8.1 p.p.m. (8H multiplet). The mass spectrum has the following relatively abundant peaks: m/e (rel. int.), 196 (25), 195 (100, M+), 166 (13), 139 (8), 102 (5), and 39 (6).

3. Discussion

Copper(I) acetylides can be prepared from ammoniacal copper(I) iodide and acetylenes.[2,4b] The generation of fresh solutions of the copper(I) salts results in a higher purity acetylide.

The substitution of aryl halides by copper(I) acetylides provides a convenient, high-yield route to aromatic acetylenes.[4] Aliphatic acetylenes can also be obtained under forcing conditions.[5] The procedure is also useful for the preparation of conjugated acetylenic ketones and alkynyl alkyl sulfides.[7] Moreover, the reaction provides the basis for a facile heterocyclic synthesis of exceedingly broad scope. Thus a halide bearing an adjacent nucleophilic substituent is substituted and cyclized by the copper(I) salt. The example described is illustrative of the preparation of indoles,[4a] benzo[b]thiophenes,[6] phthalides,[4a] benzofurans,[4a] 3-H-isobenzofurans,[7] furans,[5] 1-H-2-benzopyrans,[7] 1-H-thieno[3,4-b]-2-pyranones,[3] furo[3,2-b]pyridines,[3] furo[3,2-c]pyridines,[3] pyrrolo[3,2-b]pyridines,[7] and 4,5-dihydro-4-keto[3] benzoxepins. The furo[3,2-b]pyridine system has only been prepared by this route.[3]

1. Department of Nematology, University of California, Riverside, California 92502.
2. V. A. Sazonova, and N. Ya. Kronrod, *J. Gen. Chem. U.S.S.R.*, **26**, 2093 (1956).
3. S. A. Mladenović and C. E. Castro, *J. Heterocycl. Chem.*, **5**, 227 (1968).
4. (a) C. E. Castro, E. J. Gaughan, and D. C. Owsley, *J. Org. Chem.*, **31**, 4071 (1966); (b) C. E. Castro and R. D. Stephens, *J. Org. Chem.*, **28**, 2163, 3313 (1963); (c) S. A. Kandil and R. E. Dessy, *J. Amer. Chem. Soc.*, **88**, 3027 (1966); (d) M. D. Rausch, A. Siegel, and L. P. Klemann, *J. Org. Chem.*, **31**, 2703 (1966); (e) R. E. Atkinson, R. F. Curtis, D. M. Jones, and J. A. Taylor, *Chem. Commun.*, 718 (1967); (f) R. E. Atkinson, R. F. Curtis, and J. A. Taylor, *J. Chem. Soc., C*, 578 (1967).
5. K. Gump, S. W. Mojé, and C. E. Castro, *J. Amer. Chem. Soc.*, **89**, 6770 (1967).
6. A. M. Malte and C. E. Castro, *J. Amer. Chem. Soc.*, **89**, 6770 (1967).
7. C. E. Castro, R. H. Havlin, V. K. Honwad, A. M. Malte, and S. W. Mojé, *J. Amer. Chem. Soc.*, **91**, 6464 (1969).

2,2,3,3-TETRAMETHYLIODOCYCLOPROPANE

(Cyclopropane, 1-iodo-2,2,3,3-tetramethyl)

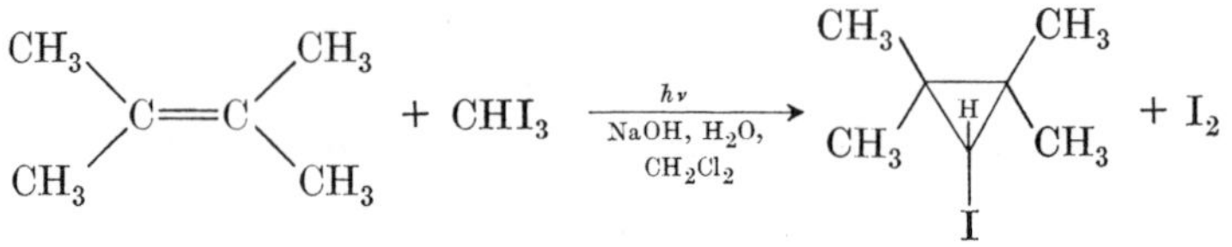

$$\text{(CH}_3\text{)}_2\text{C}=\text{C(CH}_3\text{)}_2 + \text{CHI}_3 \xrightarrow[\text{CH}_2\text{Cl}_2]{\substack{h\nu \\ \text{NaOH, H}_2\text{O}}} \text{[2,2,3,3-tetramethyl-1-iodocyclopropane]} + \text{I}_2$$

$$\text{I}_2 + 2\,\text{NaOH} \longrightarrow \text{NaI} + \text{NaOI} + \text{H}_2\text{O}$$

Submitted by T. A. MAROLEWSKI and N. C. YANG[1]
Checked by T. NAKAHIRA and K. B. WIBERG

1. Procedure

Caution! The intense emission from the light source should be shielded from visibility in order not to damage the eye-sight of the experimentalist.

In each of three 250-ml. round-bottomed Pyrex flasks are placed 8.4 g. (0.10 mole) of 2,3-dimethyl-2-butene (Note 1), 175 ml. of dichloromethane, and 50 ml. of an aqueous $5M$ sodium hydroxide solution. The flasks are kept rather full to make more efficient use of the incident light. A Teflon-covered magnetic stirring bar 2.5 cm. in length is added to each flask. Three 170 cm. by 90 cm. Pyrex crystallization dishes are almost filled with an ice-water mixture (Note 2), each dish is placed above a Mag-Mix magnetic stirrer, and each flask is then immersed in the ice-water bath with the aid of a clamp. The three assemblies are arranged symmetrically around a Hanovia quartz immersion well (No. 19434) (Note 3) cooled with running tap water containing a Hanovia 450-watt medium pressure mercury lamp (No.679A36). The edge of each flask is placed approximately 1 cm. from the wall of the well. After 2.0 g. of iodoform is added to each flask, the mixture is irradiated with stirring until the yellow color of the iodoform disappears.

This process is continued until 39.4 g. (0.10 mole) of iodoform, equally distributed between the flasks, have been consumed (Note 4). After the reaction is complete, the reaction mixtures are combined and the organic layer is separated, washed once with water, and dried over anhydrous sodium sulfate. The solvent is removed with a rotary evaporator at a water pump. The residue is transferred to a 50-ml. flask, and 1.0 g. of sodium methoxide is added (Note 5). The mixture is distilled under reduced pressure in an apparatus with a 5-cm. Vigreux side-arm (Ace Glass No. 9225). The receiver is cooled in an ice-water bath and the first fraction, which boils at 45–8° (5 mm.), n^{25} D 1.5087, is collected. The clear distillate, 14.0–15.0 g. (63–67%) yield, should be stored in a refrigerator (Notes 6 and 7).

2. Notes

1. 2,3-Dimethyl-2-butene (99%) was purchased from the Chemical Samples Co.

2. At the beginning of irradiation, the mixture is mostly ice and contains just enough water to make efficient contact with the flask. The ratio of ice to water will vary during the course of irradiation, and ice is added to replace excess water from time to time.

3. Vycor or Pyrex wells will also be satisfactory in this preparation since the irradiation is carried out in Pyrex flasks.

4. The total period of irradiation was about 8 hours; however, this may vary with the equipment used.

5. The presence of sodium methoxide is necessary to prevent the product from decomposing during the distillation.

6. The checkers also carried out the reaction using equimolar quantities of 2,3-dimethyl-2-butene and iodoform (0.1 mole each) and obtained 12.6–13.0 g. (56–58%) of the product. The submitters made the same observation. They found that the yield increased slightly as the mole ratio of olefin to iodoform was increased from 1:1 to 3:1. Use of a larger excess of olefin resulted in no further increase in yield.

7. 1-Iodo-*cis,trans*-2,3-dimethylcyclopropane, b.p. 25–27° (8–10 mm.), n^{25} D 1.5105, may be prepared in 56% yield from

trans-2-butene (Matheson Gas Products) with this procedure. Both the 2,2,3,3-tetramethyliodocyclopropane and the 1-iodo-*cis,trans*-2,3-dimethylcyclopropane prepared by this procedure give only one peak on gas chromatography. The retention times are 272 and 114 seconds, respectively, on a 60-cm. 20% SE 30 on Chromosorb W column at a temperature of 81° and a helium flow rate of 41 ml. per minute.

3. Discussion

Bromo- and iodocyclopropanes cannot be prepared by the direct halogenation of cyclopropanes. Substituted chloro- and bromocyclopropanes have been synthesized by the photochemical decomposition of α-halodiazomethanes in the presence of olefins;[2] iodocyclopropanes have been prepared from the reaction of an olefin, iodoform and potassium *t*-butoxide followed by the reduction of diiodocyclopropane formed with tri-*n*-butyl tin hydride.[3] The method described employs a readily available light source and common laboratory equipment, and is relatively safe to carry out. The method is adaptable for the preparation of bromo- and chlorocyclopropanes as well by using bromodiiodomethane or chlorodiiodomethane instead of iodoform.[4] If the olefin used will give two isomeric halocyclopropanes, the isomers are usually separable by chromatography.[4]

1. Department of Chemistry, University of Chicago, Chicago, Illinois 60637.
2. G. L. Closs and J. J. Coyle, *J. Amer. Chem. Soc.*, **87**, 4270 (1965).
3. J. P. Oliver and U. V. Rao, *J. Org. Chem.*, **31**, 2696 (1966).
4. N. C. Yang and T. A. Marolewski, *J. Amer. Chem. Soc.*, **90**, 5644 (1968).

4H-1,4-THIAZINE 1,1-DIOXIDE

[reaction scheme: 2,5-dihydrothiophene 1,1-dioxide →(1. O₃, C₂H₅OH, CH₂Cl₂, −78°; 2. SO₂)→ 2,6-diethoxy-1,4-oxathiane 4,4-dioxide →(NH₄Cl, CH₃COOH, reflux)→ 4H-1,4-thiazine 1,1-dioxide]

Submitted by Wayland E. Noland[1] and Robert D. DeMaster[2]
Checked by H. Gurien, G. Kaplan, and A. Brossi

1. Procedure

A. *cis*- and *trans*-2,6-*Diethoxy*-1,4-*oxathiane* 4,4-*Dioxide*. Ozone (Note 1) is passed into a solution of 2,5-dihydrothiophene 1,1-dioxide (30.0 g., 0.254 mole) (Note 2) in 50 ml. of absolute ethanol (Note 3) and 250 ml. of methylene chloride contained in a 1-l. three-necked round-bottomed flask fitted with a straight glass inlet tube, a calcium chloride drying tube, and a glass stopper. The solution is cooled in a dry ice-methanol bath and magnetically stirred while the ozone is being added. When the solution becomes blue (Note 4), the addition of ozone is stopped and liquid sulfur dioxide (35 ml., 0.78 mole) (Note 5) is added in portions over a period of 10–15 seconds. After 2 minutes, the cold bath is removed and the reaction solution is allowed to warm to room temperature over a period of 8–16 hours. The resulting dark brown-colored solution is poured into a 4-l. beaker containing a rapidly stirred mixture of aqueous sodium carbonate (120 g. in 1 l. of cold water) and 200 g. of ice. The reaction flask is rinsed with 50 ml. of water, which is added to the basic mixture. After being stirred for 5 minutes, the basic mixture is poured into a 2-l. separatory funnel and the lower methylene chloride layer is separated and saved. The beaker is rinsed with 200 ml. of methylene chloride and 100 ml. of water, which are then added to the separatory funnel. The contents of the separatory funnel are shaken, and the lower, methylene chloride layer is separated and saved. The aqueous layer is extracted with two more 150-ml. portions of methylene chloride. All of the methylene chloride layers and extracts are combined, and washed with 300 ml. of water and 300 ml. of saturated aqueous sodium chloride. The solution is dried over 3–6 g. of anhydrous magnesium sulfate, filtered, and evaporated with a rotary evaporator at 50–60° in a water bath under aspirator pressure. The residual cream-colored solid (50–52 g., 88–91%), m.p. 76–118°, is dissolved with magnetic stirring in 850–950 ml. of boiling heptane (Note 6) containing 1–2 g. of activated carbon and filtered hot.

The filtrate is cooled to 0° in a refrigerator overnight. The resulting precipitate is filtered, giving *cis-* and *trans-*2,6-diethoxy-1,4-oxathiane 4,4-dioxide as a white solid (42–46 g., 74–81%), m.p. 83–117° (Note 7).

B. *4H-1,4-Thiazine 1,1-Dioxide. Caution! This step should be carried out in a hood to avoid exposure to hydrogen chloride gas.*

A mixture of *cis-* and *trans-*2,6-diethoxy-1,4-oxathiane 4,4-dioxide (15.0 g., 0.0669 mole), 3.8 g. (0.071 mole) of ammonium chloride (Note 8), and 300 ml. of glacial acetic acid is placed in a 500-ml. one-necked round-bottomed flask fitted with a reflux condenser and a magnetic stirring bar. The mixture is placed in an oil bath preheated to 125–130° and refluxed, with magnetic stirring, for 25–35 minutes, during which the ammonium chloride dissolves, hydrogen chloride is evolved, and the solution becomes brownish yellow in color (Note 9). The acetic acid is evaporated with a rotary evaporator at 70–80° in a water bath under aspirator pressure. The residual yellow solid is magnetically stirred with a solution of 75 ml. of diethyl ether containing 10 ml. of isopropyl alcohol for 10 minutes (Note 10). The resulting suspension is filtered and the precipitate sucked dry on a Buchner funnel. The yellow solid (8.7–9.2 g.), m.p. 208–212°, is boiled with 225–250 ml. of isopropyl alcohol and filtered hot to remove the residual greenish black insoluble material (0.5–1 g.). The filtrate is cooled to $-10°$ to $-5°$ in a freezer overnight, causing separation of 4.6–5.3 g. (52–60%) of 4*H*-1,4-thiazine 1,1-dioxide as small yellow needles, m.p. 237–240° (Note 11), which are filtered. Concentration of the filtrate to 50 ml., followed by filtration and cooling, causes separation of an additional 1.5–2.0 g. (17–23%) of crude yellow solid, m.p. 234–240°.

2. Notes

1. A Welsbach Corporation Ozonator, style T-23, was used, with the voltage set at 120 volts and the oxygen pressure at 8 p.s.i. to give a 4–5% ozone concentration. The checkers used a Welsbach Corporation Ozonator, style T-408, to give a 1–2% ozone concentration. The input oxygen was dried by being passed through a tower of color-indicating Hammond Drierite.

2. 2,5-Dihydrothiophene 1,1-dioxide (butadiene sulfone, or 3-sulfolene) was purchased from the Aldrich Chemical Company, Inc.

3. Use of larger amounts of absolute ethanol causes formation of more of the acyclic 3-thiapentane-1,5-dial bis(diethyl acetal) 3,3-dioxide, with a corresponding reduction in yield of the cyclic product.

4. Appearance of the blue color of ozone signals complete cleavage of the double bond. Further addition of ozone could cause undesirable oxidation.

5. Sulfur dioxide was purchased in lecture-size bottles from the City Chemical Corporation. The gas was condensed into a precalibrated 50-ml. Erlenmeyer flask cooled in the dry ice-methanol bath used for cooling the ozonolysis reaction.

6. Eastman Organic Chemicals Technical Grade "Heptanes," b.p. 96–100°, containing 70% heptanes and the rest octanes, was used.

7. In one instance the submitters obtained an 85% yield when the reaction mixture was stirred with sulfur dioxide for 18 hours, followed by crystallization of the resulting crude material (32 g. per l.) without the use of charcoal.

The product is obtained as an approximately 55:45 mixture of *cis*- and *trans*-isomers, as indicated by n.m.r. absorption (CDCl$_3$ solution) at 5.33 (triplet, $J = 4$ Hz., 0.9H, CH protons of the *trans*-isomer), 4.95 (doublet of doublets, $J_{a,a} = 8$ Hz., $J_{a,e} = 2$ Hz., 1.1H, CH protons of the *cis*-isomer), 4.27–3.47 (multiplet, 4.1H, 2 CH$_3$CH_2O), 3.47–2.77 (multiplet, 4.0H, 2 CH_2SO$_2$CH_2), and 1.27 p.p.m. (triplet, $J = 7$ Hz., 5.9H, 2 CH_3). The infrared spectrum in Nujol has strong bands at 1312, 1118, 1029, and 972 cm.$^{-1}$, which are attributed to the

SO$_2$ and CO groups. The *cis*-isomer, m.p. 103–105°, can be separated from the mixture by three or four fractional crystallizations from methanol, while the *trans*-isomer, m.p. 136–137°, can be separated from the mixture (or from the residue obtained by evaporation of the methanol mother liquors from which the *cis*-isomer was crystallized) by two or three fractional crystallizations from benzene-petroleum ether (b.p. 60–68°).

8. "Baker Analyzed" Reagent Grade ammonium chloride was purchased from the J. T. Baker Chemical Company.

9. Refluxing for longer times causes formation of increased amounts of a dark greenish-brown by-product, which complicates purification by crystallization. If the acetic acid becomes black-brown, the residue (which is sometimes tarry) obtained on evaporation can be purified by a rapid chromatography through a 3.8-cm. deep column of activated alumina using acetone as a transfer agent and eluent.

10. The purpose of the wash with diethyl ether and isopropyl alcohol is to remove the remaining acetic acid and any residual hydrogen chloride, which may cause decomposition during the subsequent crystallization.

11. The analytical sample melted at 240–241.5°. The infrared spectrum in Nujol has a strong NH band at 3360, a strong band in the double bond region at 1645 and another at 1511, and a group of bands at 1265 and 1255 (medium strong) and 1238, 1226, 1102, and 1093 (all strong), some of which are attributable to the sulfonyl group, and a strong band at 692 cm.$^{-1}$. The n.m.r. spectrum (CD$_3$SOCD$_3$ solution) has an AA'BB' pattern with major peaks at 7.12 and 6.99 (2.0H) and 6.02 and 5.88 p.p.m. (2.0H), attributed to the 4 C*H* protons. The ultraviolet spectrum has maxima (95% EtOH solution) at 226 mμ (log ϵ 3.75), 230 mμ, inflection (log ϵ 3.72), 237 mμ, inflection (log ϵ 3.47), 277 mμ (log ϵ 3.52), and 287 mμ, (log ϵ 3.55).

3. Discussion

This procedure represents the first reported synthesis of *cis*- and *trans*-2,6-diethoxy-1,4-oxathiane 4,4-dioxide[3] and of its further reaction product, 4*H*-1,4-thiazine 1,1-dioxide.[3] A derivative of the latter, 3,5-diphenyl-4*H*-1,4-thiazine 1,1-dioxide, has been prepared previously by reaction of phenacyl sulfone with ammonia.[4,5] Primary amines, in addition to ammonia, can be converted to the corresponding 4-substituted 4H-1,4-thiazine 1,1-dioxides by condensation with 2,6-diethoxy-1,4-oxathiane 4,4-dioxide using the procedure described above. For example, p-aminobenzoic acid hydrochloride gave 4-(p-carboxyphenyl)-4H-1,4-thiazine 1,1-dioxide in 83% yield.[3] The submitters have also observed,[3] as have others,[4] that the 4H-1,4-thiazine 1,1-dioxide system may be N-alkylated with an alkyl halide using potassium carbonate in anhydrous acetone.

The ozonolysis reaction, followed by reductive workup with sulfur dioxide, as described in Part A of the present procedure, illustrates a general method which has been developed for the preparation of acetals.[3] Application of the procedure is illustrated by conversion of the following olefins in alcoholic solution to the corresponding acetals:[3] (1) 1-chloro-4-(o-nitrophenyl)-2-butene to o-nitrophenylacetaldehyde dimethyl acetal in 84% yield; (2) 1,4-dibromo-2-butene to bromoacetaldehyde dimethyl acetal in 67% yield; (3) 3-butenoic acid to malonaldehydic acid diethyl acetal ethyl ester in 61% yield; (4) cyclopentadiene to malonaldehyde bis(diethyl acetal) in 48% yield; and (5) 1,4-dinitro-2-butene (produced *in situ* from 1,3-butadiene and dinitrogen tetroxide) to nitroacetaldehyde diethyl acetal in 21% yield.

1. School of Chemistry, University of Minnesota, Minneapolis, Minnesota 55455.
2. Central Research Laboratory, 3M Company, St. Paul, Minnesota 55101.
3. Robert D. DeMaster, Ph.D. Dissertation, University of Minnesota, Minneapolis, Minnesota, June 1970; *Diss. Abstr. Int. B*, **31**, 5871 (1971).
4. C. R. Johnson and I. Sataty, *J. Med. Chem.*, **10**, 501 (1967).
5. I. Sataty, *J. Org. Chem.*, **34**, 250 (1969).

INDEX

This index comprises material from Volumes 50, 51, and 52 only; for previous volumes see Collective Volumes 1 through 5.

Names in small capital letters refer to the titles of individual preparations. A number in ordinary boldface type denotes the volume. A page number in boldface italics indicates that the detailed preparative directions are given or referred to; entries so treated include principal products and major by-products, special reagents or intermediates (which may or may not be isolated), compounds mentioned in the text or Notes as having been prepared by the method given, and apparatus described in detail or illustrated by a figure. Page numbers in ordinary type indicate pages on which a compound or subject is mentioned in connection with other preparations.

Acetaldehyde, directed condensation with benzophenone, **50,** 67
reaction with cyclohexylamine, **50,** 67
Acetals, synthesis, **52,** 135, 139
Acetic anhydride, with 2-heptanone to give 3-*n*-butyl-2,4-pentanedione, **51,** 90
ACETIC FORMIC ANHYDRIDE, **50,** *1*
Acetone azine, **50,** *2*
ACETONE HYDRAZONE, **50,** *2,* 28
Acetophenone, sensitizer for irradiation of bicyclo[2.2.1]hepta-2,5-diene to give quadricyclane, **51,** 133
Acetophenone N,N-dimethylhydrazone, **50,** *102*
ACETOPHENONE HYDRAZONE, **50,** *102*
3β-ACETOXY-5α-CYANOCHOLESTAN-7-ONE, **52,** *100*
1-Acetoxy-2-methylcyclohexene, **52,** *40*
p-ACETYL-α-BROMOHYDROCINNAMIC ACID, **51,** *1*
Acetyl chloride, reaction with sodium formate, **50,** 1
with propylene, aluminum chloride, and quinoline to give *trans*-3-penten-2-one, **51,** 116
2-Acetylcycloheptane-1,3-dione, **52,** *4*
2-ACETYLCYCLOPENTANE-1,3-DIONE, **52,** *1*
2-Acetylcyclopentanone, from cyclopentanone and acetic anhydride, **51,** *93*
2-Acetyl-4,4-dimethylcyclopentane-1,3-dione, **52,** *4*
2-Acetyl-5,5-dimethylcyclopentane-1,3-dione, **52,** *4*
Acetylenedicarboxylic acid, dimethyl ester, **50,** *25,* 36
Acetylenes, reaction with trimethylsilyl azide, **50,** 109
2-Acetyl-4-methylcyclopentane-1,3-dione, **52,** *4*
2-Acetylindane-1,3-dione, **52,** *4*
Acid anhydride, mixed, with sodium azide to give phenylcyclopentanecarboxylic acid azide, **51,** 48
Acrylic acid, with *p*-acetylbenzenediazonium bromide, **51,** 1
Acylation, of enol esters, **52,** 1
Alcohols, hindered, esterification, **51,** 98
primary, oxidation, **52,** 5
Aldehydes, α-phenyl-, from 2-benzyl-4,-4,6-trimethyl-5,6-dihydro-1,3(4H)-oxazine, **51,** 29
preparation, using 1,3-dithiane, **50,** 74
from primary alcohols, **52,** 5
reaction with trimethylsilylazide, **50,** 109
using acetic formic anhydride, **50,** 2
ALDEHYDES BY OXIDATION OF TERMINAL OLEFINS WITH CHROMYL CHLORIDE: 2,4,4-TRIMETHYLPENTANAL, **51,** *4*
ALDEHYDES FROM ACID CHLORIDES BY MODIFIED ROSENMUND REDUCTION: 3,4,5-TRIMETHOXYBENZALDEHYDE, **51,** *8*
ALDEHYDES FROM ACID CHLORIDES BY REDUCTION OF ESTERMESYLATES WITH SODIUM BOROHYDRIDE: CYCLOBUTANECARBOXALDEHYDE, **51,** *11*
ALDEHYDES FROM ALLYLIC ALCOHOLS AND PHENYLPALLADIUM ACETATE: 2-METHYL-3-PHENYLPROPIONALDEHYDE, **51,** *17*
ALDEHYDES FROM AROMATIC NITRILES: *p*-FORMYLBENZENESULFONAMIDE, **51,** *20*
ALDEHYDES FROM 2-BENZYL-4,4,6-

141

INDEX

TRIMETHYL-5,6-DIHYDRO-1,3-(4H)-OXAZINE: 1-PHENYLCYCLOPENTANECARBOXALDEHYDE, **51**, *24*
1-*d*-ALDEHYDES FROM ORGANOMETALLIC REAGENTS: 1-*d*-2-METHYLBUTANAL, **51**, *31*
ALDEHYDES FROM PRIMARY ALCOHOLS BY OXIDATION WITH CHROMIUM TRIOXIDE: 1-HEPTANAL, **52**, *5*
ALDEHYDES FROM *sym*-TRITHIANE: *n*-PENTADECANAL, **51**, *39*
ALDOL CONDENSATIONS, DIRECTED, **50**, 66
Alkylation, of acids, **50**, 61
  with benzyl chloromethyl ether, **52**, 17
  intramolecular to form cyclopropanes, **52**, 35
  of lithium enolates, **52**, 39
  by oxonium salts, **51**, 144
Alkylboranes, oxidation, **52**, 59
  synthesis, **52**, 59
Alkyl bromides, from alcohols, benzyl bromide, and triphenyl phosphite, **51**, 47
Alkyl chlorides, from alcohols, benzyl chloride, and triphenyl phosphite, **51**, 47
ALKYL IODIDES: NEOPENTYL IODIDE, IODOCYCLOHEXANE, **51**, *44*
Allenylacetylenes, **50**, 101
π-Allylnickel bromide, **52**, *199*
Aluminum amalgam, **52**, *78*
Aluminum chloride, with ethylene and *p*-methoxyphenylacetyl chloride to give 6-methoxy-β-tetralone, **51**, 109
  with propylene and acetyl chloride to give 4-chloropentan-2-one, **51**, 116
  with succinic anhydride and isopropenyl acetate, **52**, 1
Amine oxides, anhydrous, **50**, 55, 58
Amines, protecting group for, **50**, 12
AMINES FROM MIXED CARBOXYLIC-CARBONIC ANHYDRIDES: 1-PHENYLCYCLOPENTYLAMINE, **51**, *48*
p-Aminoacetophenone, diazotization, **51**, 1
t-Amyl iodide, from *t*-amyl alcohol, methyl iodide, and triphenyl phosphite, **51**, 47
ANDROSTAN-17β-OL, **52**, *122*
Anthracene, cyanation, **50**, 55
Arndt-Eistert reaction, modified, **50**, 77
γ-Aryl-β-diketones, general synthesis, **51**, 131
Axial alcohols, preparative methods, **50**, 15

AZIDOFORMIC ACID, *t*-BUTYL ESTER, **50**, *9*
AZIRIDINES FROM β-IODOCARBAMATES: 1,2,3,4-TETRAHYDRONAPHTHALENE(1,2)IMINE, **51**, *53*
Azoalkanes, synthesis, **52**, 11
Azocyclohexane, **52**, *15*
AZOETHANE, **52**, *11*
Azo-*n*-butane, **52**, *15*
2-Azo-2-methylpropane, **52**, *15*
Azo-*p*-nitrobenzene, **52**, *15*
Azo-*n*-propane, **52**, *15*

Benzaldehyde, by condensation of phenyllithium with 1,1,3,3-tetramethylbutyl isonitrile, **51**, 38
  by reduction of benzonitrile with Raney nickel alloy, **51**, 22
BENZALDEHYDE, 3,4,5,-TRIMETHOXY-, **51**, *8*
BENZENESULFONAMIDE, *p*-FORMYL-, **51**, *20*
Benzhydrol, **52**, *22*
BENZONITRILE, 2,4-DIMETHOXY-, **50**, *52*
Benzophenone, directed reaction with acetaldehyde, **50**, 68
Benzoylacetone, from acetophenone and acetic anhydride, **51**, *93*
BENZYL CHLOROMETHYL ETHER, **52**, *16*
2-BENZYL-2-METHYLCYCLOHEXANONE, **52**, *39*
2-BENZYL-6-METHYLCYCLOHEXANONE, **52**, *39*
3-Benzyloxy-4,5-dimethoxybenzaldehyde, by reduction of 3-benzyloxy-4,5-dimethoxybenzoyl chloride, **51**, *10*
2-Benzyl-4,4,6-trimethyl-5,6-dihydro-1,3(4H)-oxazine, from 2-methyl-2,4-pentanediol and phenylacetonitrile, **51**, *27*
BICYCLO[1.1.0]BUTANE, **51**, *55*, **52**, *32*
Bicyclo[2.2.1]hepta-2,5-diene, irradiation sensitized by acetophenone to give quadricyclane, **51**, 133
Bicyclo[2.2.0]hexa-2,5-diene, **50**, *51*
exo-Bicyclo[2.2.0]hexan-2-ol, **50**, *51*
Bicyclo[2.2.0]hex-2-ene, **50**, *51*
Bicyclo[2.2.0]hex-5-ene-2,3-dicarboxylic anhydride, **50**, *51*
BICYCLO[3.2.1]OCTAN-3-ONE, **51**, *60*
BIPHENYL, **51**, *82*
Biphenyls, unsymmetrical, **50**, 27
2,2'-Bipyridyl, use as an indicator for organolithium reagents, **52**, 39, 112
1,1-BIS(BROMETHYL)CYCLOPROPANE, **52**, *22*

INDEX

Boranes, oxidation with H_2O_2, **50**, 90; **52**, 59
2-BORENE, **51**, *66*
Boron trifluoride, with dimethyl ether and epichlorohydrin to give trimethyloxonium tetrafluoroborate, **51**, *142*
Boron trifluoride-acetic acid, with acetic anhydride and 2-heptanone to give 3-*n*-butyl-2,4-pentanedione, **51**, 90
Bromoacetaldehyde dimethyl acetal, **52**, *139*
Bromine, with 3-chlorocyclobutanecarboxylic acid and mercuric oxide to give 1-bromo-3-chlorocyclobutane, **51**, 106
1-BROMO-3-CHLOROCYCLOBUTANE, **51**, *106*
1-Bromo-3-chlorocyclobutane, with sodium to give bicyclo[1.1.0]butane, **51**, 55
3-Bromocyclobutanecarboxylic acid, **51**, 75
Bromocyclopropane, from cyclopropanecarboxylic acid, **51**, *108*
(2-Bromoethyl)benzene, **50**, *59*
3-(Bromomethyl)cyclobutyl bromide, from 3-(bromomethyl)cyclobutanecarboxylic acid, **51**, *108*
α-Bromophenylacetic acid, **50**, *31*
2-Bromothiophene, **50**, 75
1,3-BUTADIENE-1,4-DIOL *trans, trans*-DIACETATE, **50**, *24*
1,3-BUTADIENE,2,3-DIPHENYL-, **50**, *62*
BUTANAL, 1-*d*-2-METHYL-, **51**, *31*
erythro-2,3-Butanediol monomesylate, by reaction of *trans*-2-butene oxide with methanesulfonic acid, **51**, *11*
t-Butanol, with *p*-toluoyl chloride and butyllithium to give *t*-butyl *p*-toluate, **51**, 96
trans-2-Butene oxide, from *trans*-2-butene and peracetic acid, **51**, *13*
t-BUTYL AZIDOFORMATE, **50**, *9*
t-BUTYLCARBONIC DIETHYLPHOSPHORIC ANHYDRIDE, **50**, *9*
cis-4-*t*-BUTYLCYCLOHEXANOL, **50**, *13*
4-*t*-Butylcyclohexanone, **50**, 13
1-*t*-Butylcyclohexene, by reduction of *t*-butylbenzene, **50**, *92*
5-*t*-Butyl-2,3-dimethyliodobenzene, from iodine and 4-*t*-butyl-1,3-dimethylbenzene, **51**, *95*
t-Butyl hydroperoxide, **50**, 56
N-*t*-Butylhydroxylamine, **52**, *78*
t-Butyl hypochlorite, with 4-phenylurazole to give 4-phenyl-1,2,4-triazoline-3,5-dione, **51**, *123*

n-Butyllithium, **50**, 104; **52**, *19*
reaction, with 2-benzyl-4,4,6-trimethyl-5,6-dihydro-1,3-(4H)-oxazine, **51**, 25
with *sym*-trithiane and 1-bromotetradecane, **51**, 40
with 1,3-dithiane and 1-bromo-3-chloropropane to give, 5,9-dithiaspiro[3.5]nonane, **51**, 76
with *p*-toluoyl chloride and *t*-butanol to give *t*-butyl *p*-toluate, **51**, 96
sec-Butyllithium, with 1,1,3,3-tetramethylbutyl isonitrile and deuterium oxide to give N-(1-*d*-2-methylbutylidene)-1,1,3,3-tetramethylbutylamine, **51**, 33
t-BUTYLMALONIC ACID, DIETHYLESTER, **50**, *38*
3-BUTYL-2-METHYLHEPT-1-EN-3-OL, **52**, *19*
3-*n*-BUTYL-2,4-PENTANEDIONE, **51**, *90*
t-Butyl phenylacetate, from phenylacetyl chloride, *t*-butanol, and *n*-butyllithium, **51**, *98*
t-Butyl pivalate, from pivaloyl chloride and *t*-butanol, **51**, *98*
n-Butyl sulfide, with tetracyanoethylene oxide to give carbonyl cyanide, **51**, 70
t-BUTYL *p*-TOLUATE, **51**, *96*

Camphor tosylhydrazone, with methyllithium to give 2-bornene, **51**, 66
β-CARBOLINE, 1,2,3,4-TETRAHYDRO-, **51**, *136*
Carbon dioxide, anhydrous, **50**, 9
CARBONYL CYANIDE, **51**, *70*
Carbonyl cyanide, with alcohols, **51**, 72
with amines, **51**, 72
with olefins, **51**, 72
4-(*p*-Carboxyphenyl)-4H-1,4-thiazine 1,1-dioxide, **52**, *127*
o-Chlorobenzaldehyde, by reduction of *o*-chlorobenzonitrile with Raney nickel alloy in formic acid, **51**, *23*
p-Chlorobenzaldehyde, by reduction of *p*-chlorobenzonitrile with Raney nickel alloy, **51**, *22*
m-Chlorobenzoyl chloride, **50**, *16*
3-Chlorobicyclo[3.2.1]oct-2-ene, from *exo*-3,4-dichlorobicyclo[3.2.1]-oct-2-ene and lithium aluminum hydride, **51**, *61*
with sulfuric acid to give bicyclo-[3.2.1]-octan-3-one, **51**, *62*
3-CHLOROCYCLOBUTANECARBOXYLIC ACID, **51**, *73*
3-Chlorocyclobutanecarboxylic acid, with mercuric oxide and bromine to give 1-bromo-3-chlorocyclobutane, **51**, 106

INDEX

3-Chloro-1,1-cyclobutanedicarboxylic acid, from sulfuryl chloride and 1,1-cyclobutanedicarboxylic acid, **51,** *73*
4-Chloropentan-2-one, with quinoline to give *trans*-3-penten-2-one, **51,** *116*
m-CHLOROPERBENZOIC ACID, **50,** *15,* 34
3-Chloropropionitrile, **50,** *20*
Chlorosulfonyl isocyanate, in nitrile synthesis, **50,** 52
   precautions, **50,** *18*
2-Chloro-5-thiophenethiol, **50,** *106*
3α-Cholestanol, **50,** *15*
Chromium(II)-en perchlorate, **52,** *62*
Chromium(II) salts, standardization procedure for solutions, **52,** 64
Chromium trioxide-pyridine complex, **52,** *5*
Chromyl chloride, oxidation of terminal olefins, **51,** *6*
Cinnamaldehyde, by reduction of cinnamonitrile with Raney nickel alloy in formic acid, **51,** *23*
   from the ester-mesylate, **51,** *16*
Cinnamic acid, **50,** *18*
CINNAMONITRILE, **50,** *18*
Condensation, of *p*-acetylbenzenediazonium bromide with acrylic acid, **51,** *1*
Conduritol-D, **50,** *27*
Conjugate addition of Grignard reagents, **50,** *41*
CONTROLLED POTENTIAL ELECTROLYTIC REDUCTION: 1,1-BIS-(BROMOMETHYL)CYCLOPROPANE, **52,** *22*
Copper(I) chloride, use in Grignard reactions, **50,** *39*
   reaction with an organomagnesium compound, **50,** *98*
Copper(I) phenylacetylide, **52,** *128*
Coupling of acetylenes and halides, copper-promoted, **50,** *100*
Crown polyethers, complexes, **52,** *71,* *73*
   synthesis, **52,** *66*
Cuprous oxide, in thiol arylation, **50,** *75*
Curtius reaction, modification using mixed carboxylic-carbonic anhydrides, **51,** *51*
Cyanation of aromatic compounds, **50,** *53*
9-Cyanoanthracene, **50,** *55*
p-Cyanobenzenesulfonamide, reduction with Raney nickel alloy to *p*-formylbenzenesulfonamide, **51,** *20*
p-Cyano-N,N-diethylaniline, **50,** *54*
Cyanohydrins, formation by use of alkylaluminum cyanides, **52,** *96*
1-Cyano-1-hydroxy-6-methoxytetralin, **52,** *96*
Cyanomesitylene, **50,** *54*

1-CYANO-6-METHOXY-3,4-DIHYDRONAPHTHALENE, **52,** *96*
1-Cyano-2-methoxynaphthalene, **50,** *55*
4-Cyano-1-methoxynaphthalene, **50,** *55*
2-Cyanothiophene, **50,** *54*
N-Cyanovinylpyrrolidone, **50,** *54*
CYCLIC KETONES FROM 1,3-DITHIANE: CYCLOBUTANONE, **51,** *76*
Cyclobutadiene, generation *in situ,* **50,** 23
CYCLOBUTADIENEIRON TRICARBONYL, **50,** *21*
Cyclobutane, **52,** *32*
CYCLOBUTANE, 1-BROMO-3-CHLORO-, **51,** *106*
Cyclobutanecarbonyl chloride, reaction with *erythro*-2,3-butanediol monomesylate, **51,** *12*
CYCLOBUTANECARBOXALDEHYDE, **51,** *11*
CYCLOBUTANECARBOXYLIC ACID, 3-CHLORO-, **51,** *73*
1,1-Cyclobutanedicarboxylic acid, with sulfuryl chloride to give 3-chloro-1,1-cyclobutanedicarboxylic acid, **51,** *73*
CYCLOBUTANONE, **51,** *76*
CYCLOBUTENE, *cis*-3-4-DICHLORO-, **50,** *36*
2-Cyclobutyl-*cis*-4-*trans*-5-dimethyl-1,3-dioxolane, by reaction of *erythro*-3-methanesulfonyloxy-2-butyl cyclobutanecarboxylate with sodium borohydride, **51,** *12*
   hydrolysis to cyclobutanecarboxaldehyde, **51,** *13*
3,5-CYCLOHEXADIENE-1,2-DICARBOXYLIC ACID, **50,** *50*
Cyclohexane carbonitrile, **50,** *20*
1,4-Cyclohexanediol, from hydroquinone, **51,** *105*
CYCLOHEXANOL, 4-*t*-BUTYL, *cis*-, **50,** *13*
Cyclohexanol, with triphenyl phosphite and methyl iodide, **51,** *45*
CYCLOHEXANONE, 2-DIAZO-, **51,** *86*
4-CYCLOHEXENE-1,2-DICARBOXYLIC ACID, DIETHYL ESTER, *trans*-, **50,** *43*
Cyclohexylamine, **52,** *127*
   reaction with acetaldehyde, **50,** *67*
Cyclooctatretraene, chlorination, **50,** 36
   reaction with mercuric acetate, **50,** 24
CYCLOOCTENE, 1-NITRO, **50,** *84*
CYCLOPENTANECARBOXALDEHYDE, 1-PHENYL-, **51,** *24*
Cyclopentane-1,3-dione, **52,** *4*
CYCLOPENTYLAMINE, 1-PHENYL-, **51,** *48*
Cyclopropane, **52,** *32*
Cyclopropanecarboxaldehyde, by reduction of ester-mesylate, **51,** *16*

CYCLOPROPANECARBOXYLIC ACID, cis-2-PHENYL, 50, *94*
Cyclopropane derivatives, synthesis, 52, 22, 33, 132
Cyclopropenes, 50, 30

1-DECALOL, 51, *103*
2-Decalol, dehydration, 50, 92
Decarboxylation, of 3-chloro-1,1-cyclobutanedicarboxylic acid to 3-chlorocyclobutanecarboxylic acid, 51, *74*
Dehydrohalogenation, with quinoline, 51, 116
with triethylamine, 52, 36
DEHYDROXYLATION OF PHENOLS: HYDROGENOLYSIS OF PHENOLIC ETHERS: BIPHENYL, 51, *82*
"Dewar benzene," 50, *51*
trans-7,8-Diacetoxybicyclo[4.2.0]octa-2,4-diene, 50, *24*
trans, trans-1,4-DIACETOXY-1,3-BUTADIENE, 50, *24*
N,N'-Dialkylsulfamides, synthesis, 52, 11
1,2-DIAROYLCYCLOPROPANES: trans-1,2-DIBENZOYLCYCLOPROPANE, 52, *33*
1-(Diazoacetyl)naphthalene, 50, 77
16-Diazoandrost-5-ene-3β-ol-17-one, 52, *54*
2-DIAZOCYCLOALKANONES: 2-DIAZOCYCLOHEXANONE, 51, *86*
2-Diazocycloalkanones, from α-(hydroxymethylene)ketones with p-toluenesulfonyl azide, 51, 88
2-DIAZOCYCLOHEXANONE, 51, *86*
α-Diazo ketones, rearrangement, 52, 53
synthesis, 52, 53
Diazomethane, in modified Arndt-Eistert reaction, 50, 77
DIAZOMETHANE, BIS(TRIFLUOROMETHYL), 50, *6*
2-DIAZOPROPANE, 50, *5*, 27
DIBENZO-18-CROWN-6 POLYETHER, 52, *66*
trans-1,2-DIBENZOYLCYCLOPROPANE, 52, *33*
Dibenzyl sulfide, 50, *33*
Diborane, 50, 90; 52, 59
α,α'-DIBROMODIBENZYL SULFONE, 50, *31*, 65
exo-3,4-Dichlorobicyclo[3.2.1]oct-2-ene, from norbornene and ethyl trichloroacetate, 51, *60*
1,4-Dichlorobutadiene, 50, *37*
cis-3,4-DICHLOROCYCLOBUTENE, 50, 21, *36*
DICYCLOHEXYL-18-CROWN-6 POLYETHER, 52, *66*
Diels-Alder adduct, pyrolysis, 50, 37
Diels-Alder reaction, 50, 37
of 1,4-diacetoxy-1,3-butadiene, 50, 27

using 3-sulfolene, 50, 47
2,6-Diethoxy-1,4-oxathiane, 52, *135*
DIETHYLALUMINUM CYANIDE, 52, *90*
N,N-Diethylaniline, cyanation, 50, 54
DIETHYL *t*-BUTYLMALONATE, 50, *38*
Diethyl carbonate, with hydrazine hydrate to give ethyl hydrazinecarboxylate, 51, 121
Diethyl fumarate, as a dienophile, 50, 43
Diethyl isopropylidenemalonate, 50, *38*
Diethyl malonate, condensation with acetone, 50, 39
Diethyl 5-methylcoprost-3-en-3-yl phosphate, 52, *109*
Diethyl phosphorochloridate, 50, 10
reaction with metal enolates, 52, 109
N,N'-Diethylsulfamide, 52, *11*
DIETHYL trans-Δ⁴-TETRAHYDROPHTHALATE, 50, *43*
1,2-Dihydronaphthalene, with iodine isocyanate and methanol to give methyl (trans-2-iodo-1-tetralin)carbamate, 51, 112
trans-1,2-DIHYDROPHTHALIC ACID, 50, *50*
cis-1,2-Dihydrophthalic anhydride, 50, *51*
Diiododurene, from durene and iodine, 51, *95*
Diiron enneacarbonyl, 50, 21
β-DIKETONES FROM METHYL ALKYL KETONES: 3-n-BUTYL-2,4-PENTANEDIONE, 51, *90*
2,6-Dimethoxybenzaldehyde, by reduction of 2,6-dimethoxybenzonitrile with Raney nickel alloy in formic acid, 51, *23*
2,4-DIMETHOXYBENZONITRILE, 50, *52*
3,4-Dimethylbenzaldehyde, by reduction of 3,4-dimethylbenzoyl chloride, 51, *10*
4,5-DIMETHYL-1,2-BENZOQUINONE, 52, *88*
1,3-Dimethylbicyclo[1.1.0]butane, 52, *32*
N,N-DIMETHYLCYCLOHEXYLAMINE, 52, *124*
1,2-Dimethylcyclopropane, 52, *32*
N,N-Dimethyldodecylamine, 50, 56
N,N-DIMETHYLDODECYLAMINE OXIDE, 50, *56*
Dimethyl ether, with boron trifluoride diethyl etherate and epichlorohydrin to give trimethyloxonium tetrafluoroborate, 51, 142
N,N-Dimethylhydrazine, 50, 102
4,4-Dimethyl-2-neopentylpentanal, by oxidation of 4,4-dimethyl-2-neopentyl-1-pentene with chromyl

146 INDEX

chloride, **51**, *6*
2,2-Dimethyl-4-phenylbutyric acid, **50**, *58*
2,4-Dimethyl-3-sulfolene, in Diels-Alder reaction, **50**, 48
3,4-Dimethyl-3-sulfolene, in Diels-Alder reaction, **50**, 48
Dimethylsulfoxide, sodium salt, **50**, *62*
Dimethylthiocarbamyl chloride, synthesis of, **51**, *140*
with 2-naphthol to give O-2-naphthyl dimethylthiocarbamate, **51**, 139
Dinitrogen tetroxide, **50**, 84
Diphenylacetyl chloride, **52**, *36*
Diphenylacetylene, conversion to diphenylbutadiene, **50**, 63
2,3-DIPHENYL-1,3-BUTADIENE, **50**, *62*
2,2-Diphenylethanal, by oxidation of 1,1-diphenylethylene with chromyl chloride, **51**, *6*
2,2-Diphenylethyl benzoate, from 2,2-diphenylethanol, benzoyl chloride, and *n*-butyllithium, **51**, *98*
Diphenyliodonium chloride, with 2,4-pentanedione and sodium amide to give 1-phenyl-2,4-pentanedione, **51**, 128
DIPHENYLKETENE, **52**, *36*
α,α'-Diphenylthiodiglycolic acid, **50**, *31*
2,3-DIPHENYLVINYLENE SULFONE, **50**, 32, 34, *65*
Dipotassium nitrosodisulfonate, **52**, *86*, 88
Dipyridine chromium(VI) oxide, **52**, *5*
DIRECT IODINATION OF POLYALKYLBENZENES: IODODURENE, **51**, *94*
Disodium hydroxylaminedisulfonate, **52**, *83*
DISODIUM NITROSODISULFONATE, **52**, *83*
1,3-DITHIANE, **50**, *72*
1,3-Dithiane, with 1-bromo-3-chloropropane and *n*-butyllithium to give 5,9-dithiaspiro[3.5]nonane, **51**, 76
5,9-Dithiaspiro[3.5]nonane, from 1,3-dithiane, 1-bromo-3-chloropropane, and *n*-butyllithium, **51**, *76*
2,2'-DITHIENYL SULFIDE, **50**, *75*
Diynes, preparation, **50**, 101
DURENE, IODO-, **51**, *94*

Electrolytic reduction, apparatus, **52**, *23*
Enol acetates, acylation of, **52**, 1
Enol esters, preparation, **52**, 39
Epichlorohydrin, with boron trifluoride diethyl therate and dimethyl ether to give trimethyloxonium tetrafluoroborate, **51**, 142
ESTERIFICATION OF HINDERED ALCOHOLS: *t*-BUTYL *p*-TOLUATE,

51, *96*
Ethylenediamine, complexes with chromium(II) salts, **52**, 62
Ethyl diazoacetate, as source of carbethoxycarbene, **50**, 94
Ethylene, with *p*-methoxyphenylacetyl chloride and aluminum chloride to give 6-methoxy-β-tetralone, **51**, 109
Ethyl hydrazinecarboxylate, from hydrazine hydrate and diethyl carbonate, **51**, *121*
Ethylidenecyclohexylamine, **50**, *66*
Ethyl 1-iodopropionate, from ethyl 1-hydropropionate, methyl iodide, and triphenyl phosphite, **51**, *47*
ETHYL 6-METHYLPYRIDINE-2-ACETATE, **52**, *75*
Ethyl 1-naphthylacetate, **50**, 77
ETHYL PYRROLE-2-CARBOXYLATE, **51**, *100*
Ethyl trichloroacetate, with norbornene to give *exo*-3,4-dichlorobicyclo-[3.2.1]oct-2-ene, **51**, 60

FORMATION AND ALKYLATION OF SPECIFIC ENOLATE ANIONS FROM AN UNSYMMETRICAL KETONE: 2-BENZYL-2-METHYLCYCLOHEXANONE AND 2-BENZYL-6-METHYLCYCLOHEXANONE, **52**, *39*
FORMATION AND PHOTOCHEMICAL WOLFF REARRANGEMENT OF CYCLIC α-DIAZO KETONES: D-NORANDROST-5-EN-3β-OL-16-CARBOXYLIC ACIDS, **52**, *53*
FORMIC ACID, AZIDO, *t*-BUTYL ESTER, **50**, *9*
Formylation, with acetic formic anhydride, **50**, 2
p-FORMYLBENZENESULFONAMIDE, **51**, *20*
Formyl fluoride, **50**, *2*
Fremy's salt, **52**, *86*, 88

Glyoxylic acid, with tryptamine to give 1,2,3,4-tetrahydro-β-carboline, **51**, 136

1-HEPTANAL, **52**, *5*
2-Heptanone, with acetic anhydride, boron trifluoride-acetic acid, and *p*-toluenesulfonic acid to give 3-*n*-butyl-2,4-pentanedione, **51**, 90
2,4-Hexadienenitrile, **50**, *20*
Hexafluoroacetone hydrazone, **50**, *6*
HEXAFLUOROACETONE IMINE, **50**, *6*, *81*
Hexafluorothioacetone, **50**, 83
Hexamethylbicyclo[1.1.0]butane, from 1,3-dibromohexamethylcyclobu-

tane and sodium-potassium alloy, 51, *58*
Hexamethylphosphoramide, 50, 61
n-Hexanal, from 2-lithio-1,3,5-trithiane and 1-bromopentane, 51, *43*
Hunsdiecker reaction, modified; for preparation of 1-bromo-3-chlorocyclobutane, 51, 106
Hydrazine, anhydrous, 50, 3, *4, 6*
  reaction with hydrazones, 50, 102
Hydrazine hydrate, 50, 3
Hydrazoic acid, safe substitute for, 50, 107
HYDRAZONES, PREPARATION, 50, *102*
Hydroboration, of 2-methyl-2-butene, 50, 90
HYDROBORATION OF OLEFINS: (+)-ISOPINOCAMPHEOL, 52, *59*
HYDROCINNAMIC ACID, *p*-ACETYL-α-BROMO-, 51, *1*
HYDROGENATION OF AROMATIC NUCLEI: 1-DECALOL, 51, *103*
Hydrocyanation, with alkylaluminum cyanides, 52, 100
Hydrogen cyanide, reaction with triethylaluminum, 52, 90, 100
HYDROGENOLYSIS OF CARBON-HALOGEN BONDS WITH CHROMIUM(II)-EN PERCHLORATE: NAPHTHALENE FROM 1-BROMONAPHTHALENE, 52, *62*
Hydrogenolysis, of phenolic ethers to aromatics, 51, 85
  of *p*-(1-phenyl-5-tetrazolyloxy)biphenyl with palladium-on-charcoal catalyst to biphenyl, 51, 83
Hydrolysis, of 5,9-dithiaspiro[3.5]nonane to cyclobutanone, 51, 77
  of substituted *sym*-trithianes to aldehydes, 51, 42
3-Hydroxycyclohexanecarboxylic acid, from 3-hydroxybenzoic acid, 51, *105*
2-(Hydroxymethylene)cyclohexanone with *p*-toluenesulfonyl azide to give 2-diazocyclohexanone, 51, *86*

Imines of haloketones, 50, 83
Iodides, from alcohols, methyl iodide, and triphenyl phosphite, 51, 47
Iodine isocyanate, from silver isocyanate and iodine, 51, *112*
IODOCYCLOHEXANE, 51, *45*
IODODURENE, 51, *94*
trans-β-Iodoisocyanates, general synthesis from olefins with iodine isocyanate, 51, 114
Iodometric titration, 50, 17
Iridium tetrachloride, in modified Meerwein-Ponndorf reduction, 50, 13
Iron enneacarbonyl, *see* Diiron enneacarbonyl
Irradiation apparatus, 51, *133*
Irradiation, of bicyclo[2.2.1]hepta-2,5-diene to give quadricyclane, 51, 133
N-Isobutylaniline, 52, *127*
Isobutyric acid, alkylation, 50, 59
(+)-ISOPINOCAMPHEOL, 52, *59*
Isopropenyl acetate, acylation of, 52, 1

Ketones, preparation using 1,3-dithiane, 50, 74; 51, 80

Lead tetraacetate, oxidation of a hydrazone to a diazo compound, 50, 7
Lithium, reductions in amine solvents, 50, 89
Lithium aluminum hydride, reduction of *exo*-3,4-dichlorobicyclo[3.2.1]oct-2-ene to 3-chlorobicyclo[3.2.1]-oct-2-ene, 51, 61
Lithium diisopropylamide, 50, *67;* 52, *43*
Lithium dimethylcuprate, 50, *41;* 52, *109*
Lithium enolates, formation, 52, 109
Lithium enolates, preparation and alkylation, 52, 39
  reaction with diethyl phosphorochloridate, 52, 109

MACROCYCLIC POLYETHERS: DIBENZO-18-CROWN-6 POLYETHER AND DICYCLOHEXYL-18-CROWN-6 POLYETHER, 52, *66*
Malonaldehyde bis(diethyl acetal), 52, *139*
Malonaldehydic acid diethyl acetal, 52, *139*
Meerwein reaction, preparation of *p*-acetyl-α-bromohydrocinnamic acid, 51, 1
Mercuric acetate, reaction with cyclooctatetraene, 50, 24
Mercuric oxide, use in oxidation of hydrazones, 50, 28
  with 3-chlorocyclobutanecarboxylic acid and bromine to give 1-bromo-3-chlorocyclobutane, 51, 106
MERCURIC OXIDE-MODIFIED HUNSDIECKER REACTION: 1-BROMO-3-CHLOROCYCLOBUTANE, 51, *106*
Mesitylene, cyanation, 50, *54*
METALATION OF 2-METHYLPYRIDINE DERIVATIVES: ETHYL 6-METHYLPYRIDINE-2-ACETATE, 52, *75*
Methallyl alcohol, with phenylmercuric acetate to yield 2-methyl-3-phenylpropionaldehyde, 51, 17
METHALLYLBENZENE, 52, *115*
π-Methallylnickel bromide, 52, *115*
erythro-3-Methanesulfonyloxy-2-butyl

148 INDEX

cyclobutanecarboxylate, by reaction of *erythro*-2,3-butanediol monomesylate with cyclobutanecarbonyl chloride, **51,** *12*
p-Methoxybenzaldehyde, by reduction of *p*-methoxybenzonitrile with Raney nickel alloy, **51,** *22*
1-Methoxynaphthalene, cyanation, **50,** 55
2-Methoxynaphthalene, cyanation, **50,** 55
3-Methoxy-4-nitrobenzaldehyde, by reduction of 3-methoxy-4-nitrobenzoyl chloride, **51,** *10*
p-Methoxyphenylacetyl chloride, with ethylene and aluminum chloride to give 6-methoxy-β-tetralone, **51,** 109
6-METHOXY-β-TETRALONE, **51,** *109*
Methylal, **50,** 72
1-*d*-2-METHYLBUTANAL, **51,** *31*
bis-(3-Methyl-2-butyl)borane, **50,** *90*
N-(1-*d*-2-Methylbutylidene)-1,1,3,3-tetramethylbutylamine, from *sec*-butyllithium, 1,1,3,3-tetramethylbutyl isonitrile, and deuterium oxide, **51,** *33*
from *sec*-butylmagnesium bromide with 1,1,3,3-tetramethylbutyl isonitrile and deuterium oxide, **51,** *35*
5-METHYLCOPROST-3-ENE, **52,** *109*
3-Methylcyclohexene, from 2-methylcyclohexanone tosylhydrazone and methyllithium, **51,** *69*
N-Methylcyclohexylamine, **52,** *127*
3-Methylcyclopentane-1,3-dione, **52,** *4*
Methylenecyclopropanes, **50,** *30*
3-Methylheptan-4-ol, **52,** *22*
Methyl iodide, with triphenyl phosphite, and cyclohexanol, **51,** *45*
and neopentyl alcohol, **51,** *44*
METHYL (*trans*-2-IODO-1-TETRALIN)-CARBAMATE, **51,** *112*
Methyl (*trans*-2-iodo-1-tetralin)carbamate, with potassium hydroxide to give 1,2,3,4-tetrahydronaphthalene(1,2)imine, **51,** 53
Methyllithium, with camphor tosylhydrazone to give 2-bornene, **51,** 66
ether solution, **50,** *69*
standardization procedure, **50,** *69;* **52,** *46*
Methylmagnesium iodide, 1,4-addition in the presence of Cu(I), **50,** 39
2-Methyl-2-nitropropane, **52,** *78*
2-METHYL-2-NITROSOPROPANE AND ITS DIMER, **52,** *77*
Methyl D-Norandrost-5-en-3β-ol-16β-carboxylate, **52,** *56*
(S)-(-)-3-Methylpentanal, from 2-lithio-1,3,5-trithiane and (S)-(+)-1-iodo-2-methylbutane, **51,** *43*

3-Methyl-2,4-pentanedione, from 2-butanone and acetic anhydride, **51,** *93*
N-Methyl-α-phenylethylamine, **52,** *127*
2-METHYL-3-PHENYLPROPIONALDEHYDE, **51,** *17*
3-Methyl-3-phenylpropionaldehyde, from crotyl alcohol and phenylpalladium acetate, **51,** *19*
N-methylpiperidine, **52,** *127*
3-Methyl-3-sulfolene, in Diels-Alder reaction, **50,** 48

NAPHTHALENE, OCTAHYDRO-, **50,** *88*
1-NAPHTHALENEACETIC ACID, ETHYL ESTER, **50,** *77*
1-NAPHTHALENE CARBAMIC ACID, 1,2,3,4-TETRAHYDRO-2-IODO-, METHYL ESTER, **51,** *112*
Naphthalene-1-carbonitrile, **50,** *20*
2-Naphthalenecarboxyaldehyde, by reduction of 2-naphthalenecarbonitrile, **51,** *22*
NAPHTHALENE(1,2)IMINE, 1,2,3,4-TETRAHYDRO-, **51,** *53*
2-NAPHTHALENETHIOL, **51,** *139*
1-Naphthol, hydrogenation to 1-decalol, *51,* 103, 104
2-Naphthol, with dimethylthiocarbamyl chloride to give O-2-naphthyldimethylthiocarbamate, **51,** 139
1-Naphthoyl chloride, **50,** *79*
1-Naphthylacetic acid, propyl ester, **50,** *80*
O-2-Naphthyl dimethylthiocarbamate, from 2-naphthol and dimethylthiocarbamyl chloride, **51,** *139*
thermolysis to S-2-naphthyl dimethylthiocarbamate, **51,** 140
S-2-Naphthyl dimethylthiocarbamate, hydrolysis with potassium hydroxide to 2-naphthalenethiol, **51,** 140
Neopentyl alcohol, with triphenylphosphite and methyl iodide, **51,** 44
NEOPENTYL IODIDE, **51,** *44*
Nickel carbonyl, precautions for handling, **52,** *117*
reaction with allyl halides, **52,** 115
Nitriles, from carboxylic acids, **50,** 20
Nitroacetaldehyde diethyl acetal, **52,** *139*
Nitro compounds, preparation, **50,** 88
1-NITROCYCLOOCTENE, **50,** *84*
Nitrogen atmosphere, apparatus for maintaining, **52,** *46*
Nitrogen, purification, **50,** 69
1-Nitro-1-octadecene, **50,** *86*
o-Nitrophenylacetaldehyde dimethyl acetal, **52,** *139*
4-*p*-Nitrophenyl-1,2,4-triazoline-3,5-dione, synthesis of, **51,** *125*

Nitrosation, of ketones, **52**, 53
Nitroso compounds, synthesis, **52**, 77
Nonan-5-ol, **52**, *22*
D-NORANDROST-5-EN-3β-OL-16-CARBOXYLIC ACIDS, **52**, *53*
Norbornene, with ethyl trichloroacetate to give *exo*-3,4-dichlorobicyclo-[3.2.1]oct-2-ene, **51**, *60*

$\Delta^{1,9}$-Octalin, **50**, *89*
$\Delta^{1,10}$OCTALIN, **50**, *88*
n-Octanal, from 2-lithio-1,3,5-trithiane and 1-bromoheptane, **51**, *43*
Olefins, from tosylhydrazones and methyllithium, **51**, 69
  synthesis, **52**, 109, 115
  terminal, with chromyl chloride, **51**, 6
Organolithium reagents, preparation, **52**, 21
  standardization procedure, **52**, 46
Oxidation, of primary alcohols to aldehydes, **52**, 5
  with chromium trioxide-pyridine complex, **52**, 5
  with hydrogen peroxide, **52**, 59
  with the nitrosodisulfonate radical, **52**, 83, 88
  with ozone, **52**, 135
  with potassium permanganate, **52**, 77
  with sodium hypobromite, **52**, 77
  with sodium hypochlorite, **52**, 11
  of terminal olefins with chromyl chloride, **51**, 6
  of 2,4,4-trimethyl-1-pentene with chromyl chloride, **51**, 4
OXIDATION WITH THE NITROSODISULFONATE RADICAL. I. PREPARATION AND USE OF DISODIUM NITROSODISULFONATE: TRIMETHYL-*p*-BENZOQUINONE, **52**, *83*
OXIDATION WITH THE NITROSODISULFONATE RADICAL. II. USE OF DIPOTASSIUM NITROSODISULFONATE (FREMY'S SALT): 4,5-DIMETHYL-1,2-BENZOQUINONE, **52**, *88*
Oximes, preparation, **50**, 88
Oximino ketones, synthesis, **52**, 53
16-Oximinoandrost-5-en-3β-ol-17-one, **52**, *53*
Oxygen, analysis for active, **50**, 16
Ozonides, reduction with sulfur dioxide, **52**, 135

Palladium-on-charcoal catalyst, biphenyl from *p*-(1-phenyl-5-tetrazolyloxy)-biphenyl and hydrogen, **51**, 83
n-PENTADECANAL, **51**, *39*
n-Pentadecanal dimethyl acetal, by methanolysis of 2-tetradecyl-*sym*-trithiane, **51**, *40*

1,3-PENTADIYNE, 1-PHENYL, **50**, *97*
1,4-PENTADIYNE, 1-PHENYL, **50**, *97*
n-Pentanal, by condensation of butyllithium with 1,1,3,3-tetramethylbutyl isonitrile, **51**, *38*
2,4-Pentanedione, 3-alkyl, **51**, *93*
2,4-PENTANEDIONE, 3-*n*-BUTYL, **51**, *90*
2,4-PENTANEDIONE, 1-PHENYL-, **51**, *128*
2,4-Pentanedione, with sodium amide and diphenyliodonium chloride to give 1-phenyl-2,4-pentanedione, **51**, 128
trans-3-PENTEN-2-ONE, **51**, *115*
3-Pentanol, **52**, *22*
Pent-1-en-3-ol, **52**, *22*
PERBENZOIC ACID, *m*-CHLORO, **50**, *15*
Periodic acid dihydrate, with iodine and durene to give iododurene, **51**, 94
Phenylacetaldehyde, from 2-lithio-1,3,5-trithiane and benzyl bromide, **51**, *43*
Phenylacetic acid, bromination, **50**, 31
Phenylacetonitrile, **50**, *20*
Phenylacetylene, reaction with ethyl magnesium bromide, **50**, 98
Phenylation, of β-diketones with diphenyliodonium chloride, **51**, 131
  of ketoesters with diphenyliodonium chloride, **51**, 131
  of nitroalkanes with diphenyliodonium chloride, **51**, 132
PHENYLATION WITH DIPHENYLIODONIUM CHLORIDE: 1-PHENYL-2,4-PENTANEDIONE, **51**, *128*
4-Phenyl-1-carbethoxysemicarbazide, from ethyl hydrazinecarboxylate and phenyl isocyanate, **51**, *122*
  with potassium hydroxide to give 4-phenylurazole, **51**, 122
1-Phenyl-5-chlorotetrazole, with *p*-phenylphenol to give *p*-(1-phenyl-5-tetrazolyloxy)biphenyl, **51**, 82
β-PHENYLCINNAMALDEHYDE, **50**, *65*
1-PHENYLCYCLOPENTANECARBOXALDEHYDE, **51**, *24*
1-Phenylcyclopentanecarboxylic acid, with ethyl chlorocarbonate to give mixed carboxylic-carbonic anhydride, **51**, 48
1-PHENYLCYCLOPENTYLAMINE, **51**, *48*
Phenylcyclopentylamine, by hydrolysis of phenylcyclopentyl isocyanate, **51**, *49*
Phenylcyclopentyl isocyanate, by thermolysis of phenylcyclopentanecarboxylic acid azide, **51**, *49*
2-(1-Phenylcyclopentyl)-4,4,6-trimethyl-

5,6-dihydro-1,3(4H)-oxazine, from 2-benzyl-4,4,6-trimethyl-5,6-dihydro-1,3(4H)-oxazine, 1,4-dibromobutane, and n-butyllithium, **51**, *24*

2-(1-Phenylcyclopentyl)-4,4,6-trimethyltetrahydro-1,3-oxazine, by reduction of 2-(1-phenylcyclopentyl)-4,4,6-trimethyl-5,6-dihydro-1,3-(4H)-oxazine with sodium borohydride, **51**, *25*

hydrolysis, to 1-phenylcyclopentanecarboxaldehyde, **51**, *26*

α-Phenylcyclopropane, **52**, *32*

Phenylcyclopropanecarboxaldehyde, from 2-benzyl-4,4,6-trimethyl-5,6-dihydro-1,3(4H)-oxazine, **51**, *29*

cis-2-PHENYLCYCLOPROPANECARBOXYLIC ACID, **50**, *94*

trans-2-Phenylcyclopropanecarboxylic acid, **50**, *96*

α-Phenylethylamine, **52**, *127*

2-Phenylethyl iodide, from 2-phenylethanol, methyl iodide, and triphenyl phosphite, **51**, *47*

Phenylethynylmagnesium bromide, **50**, *97*

Phenyl isocyanate, with ethyl hydrazinecarboxylate to give 4-phenyl-1-carbethoxysemicarbazide, **51**, *122*

Phenylmercuric acetate, with methallyl alcohol to yield 2-methyl-3-phenylpropionaldehyde, **51**, *17*

1-PHENYL-1,3-PENTADIYNE, **50**, *97*

1-PHENYL-1,4-PENTADIYNE, **50**, *97*

α-Phenylpentanal, from 2-benzyl-4,4,6-trimethyl-5,6-dihydro-1,3(4H)-oxazine, **51**, *29*

1-PHENYL-2,4-PENTANEDIONE, **51**, *128*

3-Phenyl-2,4-pentanedione, from phenylacetone and acetic anhydride, **51**, *93*

p-Phenylphenol, with 1-phenyl-5-chlorotetrazole to give phenolic ether, **51**, 82

2-Phenylpropanal, by oxidation of 2-phenylpropene with chromyl chloride, **51**, *6*

3-Phenylpropanal, from allyl alcohol and phenylpalladium acetate, **51**, *19*

2-PHENYL[3,2-b]PYRIDINE, **52**, *128*

p-(1-Phenyl-5-tetrazolyloxy)biphenyl, from p-phenylphenol and 1-phenyl-5-chlorotetrazole, **51**, *82*

hydrogenation to biphenyl, **51**, 83

4-PHENYL-1,2,4-TRIAZOLINE-3,5-DIONE, **51**, *121*

4-Phenyl-1,2,4-triazoline-3,5-dione, reactions of, **51**, 126

4-Phenylurazole, from 4-phenyl-1-carbethoxysemicarbazide and potassium hydroxide, **51**, *122*

with t-butyl hypochlorite to give 4-phenyl-1,2,4-triazoline-3,5-dione, **51**, 123

Phosphinimines, **50**, 109

Photochemical reactions, dissociation of trihalomethanes, **52**, 132
rearrangement, **52**, 53

Phthalic acid, reduction, **50**, 50

Pivalaldehyde, by condensation of t-butylmagnesium bromide with 1,1,3,3-tetramethylbutyl isonitrile, **51**, 38

by reduction of ester-mesylate, **51**, *16*

Pivalonitrile, **50**, *20*

Polyalkylbenzenes, with iodine to give iodo derivatives, **51**, 95

Potassium acetate complex with dicyclohexyl-18-crown-6 polyether, **52**, 71

Potassium amide, **52**, 75

Potassium azide, **50**, 10

Potassium t-butoxide, **52**, *53*

Potassium hydroxide complex with dicyclohexyl-18-crown-6 polyether, **52**, *71*

PREPARATION OF CYANO COMPOUNDS USING ALKYLALUMINUM INTERMEDIATES. I. DIETHYLALUMINUM CYANIDE, **52**, *90*

PREPARATION OF CYANO COMPOUNDS USING ALKYLALUMINUM INTERMEDIATES. II. 1-CYANO-6-METHOXY-3,4-DIHYDRONAPHTHALENE, **52**, *96*

PREPARATION OF CYANO COMPOUNDS USING ALKYLALUMINUM INTERMEDIATES. III. 3β-ACETOXY-5α-CYANOCHOLESTAN-7-ONE, **52**, *100*

PREPARATION AND REDUCTIVE CLEAVAGE OF ENOL PHOSPHATES: 5-METHYLCOPROST-3-ENE, **52**, *109*

1,3-Propanedithiol, **50**, 72

Propargyl bromide, coupling with an organocopper reagent, **50**, 98

PROPIONALDEHYDE, 2-METHYL-3-PHENYL-, **51**, *17*

Propylene, with acetyl chloride, aluminum chloride, and quinoline to give trans-3-penten-2-one, **51**, 115

with acetyl chloride and aluminum chloride to give 4-chloropentan-2-one, **51**, 116

α-Pyrone, irradiation of, **50**, 23

Pyrrole, with trichloroacetyl chloride to give pyrrol-2-yl trichloromethyl ketone, **51**, 100

PYRROLE-2-CARBOXYLIC ACID, ETHYL ESTER, **51**, *100*

INDEX

Pyrrole-2-carboxylic acid esters, from pyrrol-2-yl trichloromethyl ketone, **51**, 102
Pyrrol-2-yl-trichloromethyl ketone, with ethanol to give ethyl pyrrole-2-carboxylate, **51**, 100

QUADRICYCLANE, **51**, *133*
Quadricyclane, preparation of substituted derivatives, **51**, 135
reactions of, **51**, 135
Quinoline, with 4-chloropentan-2-one to give *trans*-3-penten-2-one, **51**, 116

Raney nickel alloy, reduction of aromatic nitriles to aldehydes, **51**, 22
REACTION OF ARYL HALIDES WITH π-ALLYLNICKEL HALIDES: METHALLYLBENZENE, **52**, *115*
Reduction, of acid chlorides with palladium-on-carbon catalyst to give aldehydes, **51**, 10
with aluminum amalgam, **52**, 77
of aromatic nuclei, **51**, 105
with Chromium(II) salts, **52**, 62
by controlled-potential electrolysis, **52**, 22
of *p*-cyanobenzenesulfonamide with Raney nickel alloy to *p*-formylbenzenesulfonamide, **51**, 20
with hydroxylamine, **52**, 128
by lithium aluminum hydride of *exo*-3,4-dichlorobicyclo[3.2.1]oct-2-ene to 3-chlorobicyclo[3.2.1]oct-2-ene, **51**, 61
by sodium borohydride of *erythro*-3-methanesulfonyloxy-2-butyl cyclobutanecarboxylate, **51**, 12
with sodium borohydride, **52**, 122
by sodium borohydride of 2-(1-phenylcyclopentyl)-4,4,6-trimethyl-5,6-dihydro-1,3(4H)-oxazine, **51**, 25
with sodium cyanoborohydride, **52**, 124
with sulfur dioxide, **52**, 83, 135
REDUCTION OF KETONES BY USE OF THE TOSYLHYDRAZONE DERIVATIVES: ANDROSTAN-17β-OL, **52**, *122*
REDUCTIVE AMINATION WITH SODIUM CYANOBOROHYDRIDE: N,N-DIMETHYLCYCLOHEXYL-AMINE, **52**, *124*
Reference electrode for electrolytic reduction, **52**, 28
Resorcinol dimethyl ether, **50**, 52
Rhodium-on-alumina, catalyzed reduction of aromatic nuclei, **51**, 105
Rosemund reduction, 3,4,5-trimethoxybenzaldehyde, **51**, 8

Sebacic acid dinitrile, **50**, *20*

Shikimic acid, **50**, *27*
Silver benzoate, as catalyst in decomposition of diazoketones, **50**, 78
Silver isocyanate, with iodine to give iodine isocyanate, **51**, 112
Sodium, with 1-bromo-3-chlorocyclobutane to give bicyclo[1.1.0]butane, **51**, 55
Sodium amalgam, **50**, 50, *51*
Sodium amide, with 2,4-pentanedione and diphenyliodonium chloride to give 1-phenyl-2,4-pentanedione, **51**, 128
Sodium azide, **50**, 107
with mixed carboxylic-carbonic anhydrides, **51**, 49
Sodium borohydride, reduction of *erythro*-3-methanesulfonyloxy-2-butyl cyclobutanecarboxylate, **51**, 12
reduction of 2-(1-phenylcyclopentyl)-4,4,6-trimethyl-5,6-dihydro-1,3-(4H)-oxazine to 2-(1-phenylcyclopentyl)-4,4,6-trimethyltetrahydro-1,3-oxazine, **51**, 25
Sodium cyanoborohydride, use in reductions, **52**, 124
Sodium formate, reaction with acetyl chloride, **50**, 1
Sommelet reaction, **50**, 71
Spiropentane, **52**, *32*
from pentaerythrityltetrabromide and sodium, **51**, *58*
Steam distillation, of volatile aldehydes, **51**, 33, 36
Styrene, reaction with carbethoxycarbene, **50**, 94
SUBSTITUTION OF ARYL HALIDES WITH COPPER(I) ACETYLIDES: 2-PHENYL[3,2-b]PYRIDINE, **52**, *128*
Succinic acid mononitrile, ethyl ester, **50**, *20*
Succinic anhydride, **52**, 1
Sulfides, aromatic, **50**, 76
3-Sulfolene, as a source of 1,3-butadiene *in situ*, **50**, 43
Sulfones, bromination, **50**, 31
Sulfur, reaction with organo-lithium compounds, **50**, 105
Sulfuryl chloride, with 1,1-cyclobutanedicarboxylic acid to give 3-chloro-1,1-cyclobutanedicarboxylic acid, **51**, 73

Tetracyanoethylene oxide, with *n*-butyl sulfide to give carbonyl cyanide, **51**, 70
2-Tetradecyl-*sym*-trithiane, by reaction of 1-bromotetradecane with *sym*-trithiane in presence of *n*-butyllithium, **51**, 39

INDEX

Tetraethylammonium tetrafluoroborate, 52, 29
1,2,3,4-TETRAHYDRO-β-CARBOLINE, 51, 136
1,2,3,4-Tetrahydro-β-carboline, synthesis of substituted derivatives, 51, 138
1,2,3,4-TETRAHYDRONAPHTHALENE(1,2)IMINE, 51, 53
β-TETRALONE, 6-METHOXY-, 51, 109
β-Tetralones, general synthesis of substituted derivatives, 51, 111
1,1,3,3-Tetramethylbutyl isonitrile, from N-(1,1,3,3-tetramethylbutyl)-formamide and thionyl chloride, 51, 31
2,4,4,6-Tetramethyl-5,6-dihydro-1,3-(4H)-oxazine, for synthesis of substituted acetaldehydes, 51, 30
2,2,3,3-TETRAMETHYLIODOCYCLOPROPANE, 52, 132
Thermolysis, 1-phenylcyclopentanecarboxylic acid azide to 1-phenylcyclopentyl isocyanate, 51, 49
4H-1,4-THIAZINE 1,1-DIOXIDE, 52, 135
2-Thienyllithium, 50, 104
THIIRENE 1,1-DIOXIDE, DIPHENYL, 50, 65
2,2'-THIODITHIOPHENE, 50, 75
Thiols, general synthetic method, 50, 106
Thiophene, cyanation, 50, 54
2-THIOPHENETHIOL, 50, 75, 104
3-Thiophenethiol, 50, 106
THIOPHENOLS FROM PHENOLS: 2-NAPHTHALENETHIOL, 51, 139
o-Tolualdehyde, by reduction of o-tolunitrile with Raney nickel alloy in formic acid, 51, 23
p-Toluenesulfonyl azide, with 2-(hydroxymethylene)cyclohexanone to give 2-diazocyclohexanone, 51, 86
p-TOLUIC ACID, t-BUTYL ESTER, 51, 96
p-Toluoyl chloride, with t-butanol and n-butyllithium to give t-butyl p-toluate, 51, 96
Tosylhydrazones, formation, 52, 122
with methyllithium to give olefins, 51, 69
reduction, 52, 122
Trialkyloxonium salts, as alkylating agents, 51, 144
Triazoles, general route to, 50, 109

1,2,4-TRIAZOLINE-3,5-DIONE, 4-PHENYL-, 51, 121
Tri-n-butylcarbinol, 52, 22
Trichloroacetyl chloride, with pyrrole to give pyrrol-2-yl trichloromethyl ketone, 51, 100
Triethylaluminum, apparatus and procedures for handling, 52, 90, 96, 100
Triethylamine, in synthesis of diazoketones, 50, 77
bis(Trifluoromethyl)carbene, 50, 8
BIS(TRIFLUOROMETHYL)DIAZOMETHANE, 50, 6
3,4,5-TRIMETHOXYBENZALDEHYDE, 51, 8
3,4,5-Trimethoxybenzoyl chloride, reduction to 3,4,5-trimethoxybenzaldehyde, 51, 8
TRIMETHYL-p-BENZOQUINONE, 52, 83
Trimethylchlorosilane, 50, 107
Trimethylcyclohexanones, reduction to axial alcohols, 50, 15
TRIMETHYLOXONIUM TETRAFLUOROBORATE, 51, 142
Trimethyloxonium tetrafluoroborate, reactions of, 51, 144
2,4,4-TRIMETHYLPENTANAL, 51, 4
TRIMETHYLSILYL AZIDE, 50, 107
Triphenylphosphine imine, 50, 83
Triphenyl phosphite, with methyl iodide and cyclohexanol, 51, 45
with neopentyl alcohol and methyl iodide, 51, 44
sym-Trithiane, reaction with 1-bromotetradecane in presence of n-butyllithium, 51, 39
Tryptamine, with glyoxylic acid to give 1,2,3,4-tetrahydro-β-carboline, 51, 136

α,β,-Unsaturated carbonyl compounds, preparative method, 50, 70

Vacuum manifold system, 51, 56
Vanadium oxyacetylacetonate, 50, 56
N-Vinylpyrrolidone, cyanation, 50, 54

Wurtz reaction, bicyclo[1.1.0]butane from 1-bromo-3-chlorocyclobutane, 51, 55

Zinc, cyclopropane from 1,3-dichloropropane, 51, 58

Unchecked Procedures

Received during the period July 1, 1971-June 30, 1972

In accordance with a policy adopted by the Board of Editors beginning with Volume 50, which, as noted in the Editor's Preface, is intended to make procedures available more rapidly, procedures received by the Secretary during the year, whether subsequently accepted, modified, or rejected for publication in Organic Syntheses, will be made available for purchase at the price of $2 per procedure, prepaid upon request to the Secretary:

> Dr. Wayland E. Noland, Secretary
> Organic Syntheses
> School of Chemistry - Smith Hall
> University of Minnesota
> Minneapolis, Minnesota 55455

Payment must accompany the order, and should be made payable to Organic Syntheses, Inc. (not to the Secretary). Purchase orders not accompanied by payment will not be accepted. Procedures may be ordered by number and/or title from the list which follows.

It should be emphasized that the procedures which are being made available are unedited and have been reproduced just as they are received from the submitters. The procedures have not been checked in the form available, and many of them have been or will be rejected for publication in Organic Syntheses either before or after the checking process. For this reason, Organic Syntheses can provide no assurance whatever that the procedures will work as described, and offers no comment as to what safety hazards may be involved. Consequently, more than usual caution should be employed in following the directions in the procedures.

Organic Syntheses welcomes, on a strictly voluntary basis, comments from persons who attempt to carry out the procedures. For this purpose, a Checker's Report form will be mailed out with each unchecked procedure ordered. Procedures which have been checked by or under the supervision of a member of the Board of Editors will continue to be published in the volumes of Organic Syntheses, as in the past. It is anticipated that some of the procedures in the list will be published (often in revised form) in Organic Syntheses in future volumes.

-Wayland E. Noland

1770   3-Butyl-2-methyl-1-hepten-3-ol

P. J. Pearce, D. H. Richards, and N. F. Scilly, Non-Metallic
Materials Branch, Explosives Research and Development Establishment,
Ministry of Defence, Waltham Abbey, Essex, England

$$4\ Li + 2\ CH_3CH_2CH_2CH_2Br + CH_2=\underset{\underset{CH_3}{|}}{C}-\underset{\underset{O}{||}}{C}OCH_3 \xrightarrow[-20°]{THF} \xrightarrow[H_2O]{HCl}$$

$$CH_2=\underset{\underset{CH_3}{|}}{C}-\underset{\underset{OH}{|}}{\overset{\overset{CH_3CH_2CH_2CH_3}{|}}{C}}-CH_2CH_2CH_2CH_3$$

80% (>99% pure by v.p.c.)

1771   Cycloundecanone

J. Wohllebe and E. W. Garbisch, Jr., Department of Chemistry,
University of Minnesota, Minneapolis, Minnesota   55455

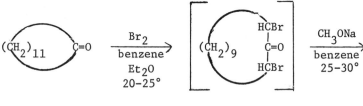

91-93%   91%
                                              (83-85% overall)

1772 Tetrahydro-4H-pyran-4-one

G. R. Owen and C. B. Reese, University Chemical Laboratory, Lensfield Road, Cambridge, CB2 1EW, England

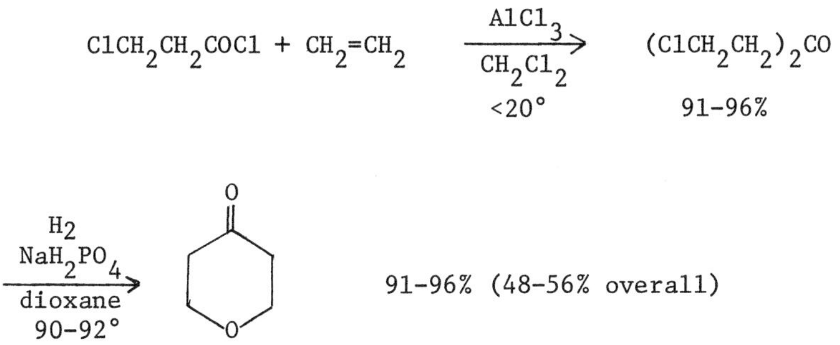

$$ClCH_2CH_2COCl + CH_2=CH_2 \xrightarrow[CH_2Cl_2]{AlCl_3} (ClCH_2CH_2)_2CO$$
$$<20° \quad 91-96\%$$

$$\xrightarrow[\text{dioxane}]{H_2 \atop NaH_2PO_4} \quad \text{[tetrahydropyran-4-one]} \quad 91-96\% \text{ (48-56\% overall)}$$
$$90-92°$$

1773 Quinoline-4-carbonitrile

S. D. Jolad and S. Rajagopal, Department of Chemistry, Karnatak University, Dharwar, Mysore State, India

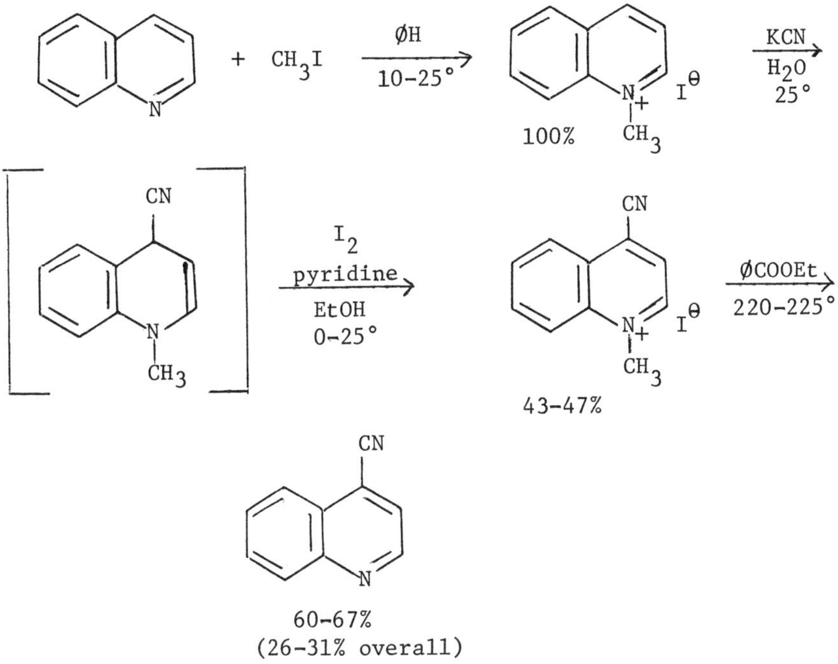

[quinoline] + CH$_3$I $\xrightarrow[10-25°]{\varnothing H}$ [N-methylquinolinium] I$^\ominus$ $\xrightarrow[25°]{KCN \atop H_2O}$

100%

[1,4-dihydro-4-cyano-1-methylquinoline] $\xrightarrow[\text{EtOH} \atop 0-25°]{I_2 \atop \text{pyridine}}$ [4-cyano-1-methylquinolinium] I$^\ominus$ $\xrightarrow[220-225°]{\varnothing COOEt}$

43-47%

[quinoline-4-carbonitrile]

60-67%
(26-31% overall)

Ketones and Alcohols from Organoboranes. (1) Phenyl Heptyl Ketone. (2) 1-Hexanol. (3) 1-Octanol

Hiromichi Kono and John Hooz, Department of Chemistry, University of Alberta, Edmonton, Alberta, Canada

$$CH_3(CH_2)_3CH=CH_2 + BH_3 \xrightarrow[20-25°]{N_2 \atop THF} [(C_6H_{13})_3B]$$

$$\xrightarrow[THF, reflux]{N_2CHCOC_6H_5} \xrightarrow[AcONa \atop <20°]{H_2O_2} \underline{n}\text{-}C_7H_{15}COC_6H_5$$
75-80%

$$[(C_6H_{13})_3B] \xrightarrow[THF \atop 35°]{NaOH \atop H_2O_2} \underline{n}\text{-}C_6H_{13}OH$$
69-76%
(>95% purity)

$$(CH_3)_2C=CHCH_3 + BH_3 \xrightarrow[<10°]{N_2 \atop THF} [[(CH_3)_2\overset{CH_3}{\overset{|}{CH}}CH]_2BH]$$

$$\xrightarrow[THF, 20-25°]{\underline{n}\text{-}C_6H_{13}CH=CH_2} [[(CH_3)_2\overset{CH_3}{\overset{|}{CH}}CH]_2B\text{-}\underline{n}\text{-}C_8H_{17}]$$

$$\xrightarrow[THF \atop 30-35°]{NaOH \atop H_2O_2} (CH_3)_2CHCHCH_3 + \underline{n}\text{-}C_8H_{17}OH$$
OH
57-59%$$77-82%

1775 Ketones from Carboxylic Acid Chlorides and Organocopper Reagents: Methyl 4-Oxoöctanoate

Gary H. Posner and Charles E. Whitten, Department of Chemistry, The Johns Hopkins University, Baltimore, Maryland 21218

$$CuI + 2\ \underline{n}\text{-BuLi} \xrightarrow[\substack{N_2 \\ -30°}]{Et_2O} (\underline{n}\text{-Bu})_2CuLi$$
in pentane

$$CH_3OCO(CH_2)_2COCl + (\underline{n}\text{-Bu})_2CuLi \xrightarrow[\substack{N_2 \\ -78°}]{Et_2O} CH_3OCO(CH_2)_2CO(CH_2)_3CH_3$$

1776 Diamantane (Congressane)

Tamara M. Gund, Wilfried Thielecke, and Paul v.R. Schleyer, Dept. of Chemistry, Frick Chemical Laboratory, Princeton University, Princeton, New Jersey

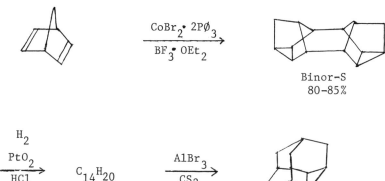

Binor-S
80-85%

$$\xrightarrow[\text{AcOH}]{\substack{H_2 \\ PtO_2 \\ HCl}} C_{14}H_{20} \xrightarrow[\substack{\text{or} \\ C_6H_{12}}]{AlBr_3 \\ CS_2}$$

Tetrahydro-Binor-S
90-97%

Diamantane
62-71%
(45-59% overall)

1777   Cyclopropanes via Abnormal Hofmann Elimination: Ethyl 1-Benzylcyclopropanecarboxylate

Carl Kaiser and Joseph Weinstock, Research and Development Division, Smith Kline and French Laboratories, 1500 Spring Garden Street, Philadelphia, Pa. 19101

$$ClCH_2CH_2N(CH_3)_2 \cdot HCl \xrightarrow[H_2O]{K_2CO_3} ClCH_2CH_2N(CH_3)_2$$

$$\emptyset CH_2CH(COOEt)_2 + ClCH_2CH_2N(CH_3)_2 \xrightarrow[25-100°]{\substack{NaH \\ (CH_3)_2SO}} \left[\begin{array}{c} \emptyset CH_2C(COOEt)_2 \\ | \\ CH_2CH_2N(CH_3)_2 \end{array} \right] \xrightarrow[25°]{CH_3I} \left[\begin{array}{c} \emptyset CH_2C(COOEt)_2 \\ | \\ CH_2CH_2\overset{+}{N}(CH_3)_3 \\ I^{\ominus} \end{array} \right] \xrightarrow[\substack{\text{anion} \\ \text{exchange} \\ \text{resin}}]{OH^{\ominus}} \xrightarrow[25°]{EtOH}$$

$$\left[\begin{array}{c} \emptyset CH_2C(COOEt)_2 \\ | \\ CH_2CH_2\overset{+}{N}(CH_3)_3 \\ EtO^{\ominus} \end{array} \right] \xrightarrow[150-200°]{\substack{-CO(OEt)_2 \\ -N(CH_3)_3}} \emptyset CH_2\underset{CH_2}{\overset{COOEt}{\underset{|}{C}}}\!\!-\!\!CH_2$$

65-75%

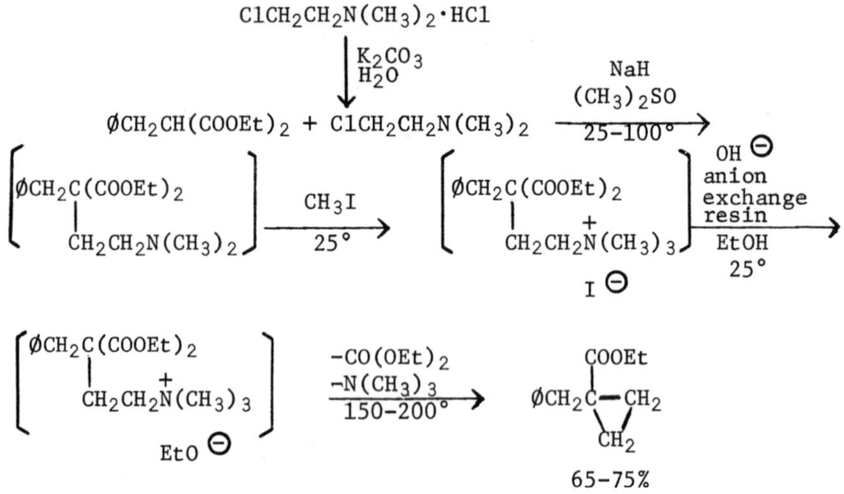

1778   Olefins via Hofmann Elimination: Use of Anion Exchange Resin for Preparation of Quaternary Ammonium Alkoxides

Carl Kaiser and Joseph Weinstock, Research and Development Division, Smith Kline and French Laboratories, 1500 Spring Garden Street, Philadelphia, Pa. 19101

$$\emptyset_2CHOCH_2CH_2N(CH_3)_2 + CH_3I \xrightarrow[25°]{\text{acetone}} \emptyset_2CHOCH_2CH_2\overset{+}{N}(CH_3)_3 \; I^{\ominus}$$

80-95%

$$\xrightarrow[\substack{CH_3OH \\ 25°}]{\substack{OH^{\ominus} \\ \text{anion} \\ \text{exchange} \\ \text{resin}}} \left[\emptyset_2CHOCH_2CH_2\overset{+}{N}(CH_3)_3 \; CH_3O^{\ominus} \right] \xrightarrow[100°]{\substack{-CH_3OH \\ -N(CH_3)_3}} \emptyset_2CHOCH=CH_2$$

79-89%
(70-78% overall)

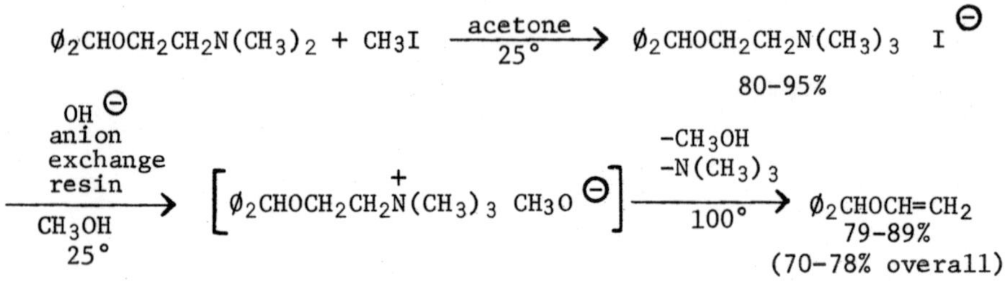

1779   Tellurophene

Francesco Fringuelli and Aldo Taticchi, Instituto di Chimica Organica, Universita di Perugia, via Elce die Sotto 10-06100, Perugia, Italy

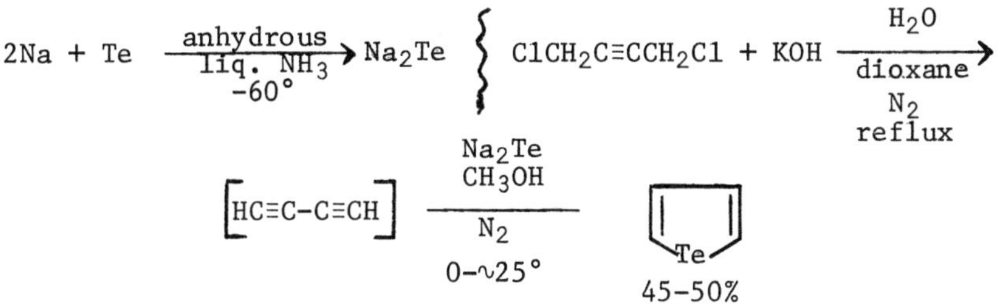

$$2Na + Te \xrightarrow[-60°]{anhydrous\ liq.\ NH_3} Na_2Te \begin{cases} ClCH_2C\equiv CCH_2Cl + KOH \xrightarrow[N_2]{H_2O\ dioxane\ reflux} \\ [HC\equiv C-C\equiv CH] \xrightarrow[N_2\ 0-\sim 25°]{Na_2Te\ CH_3OH} \underset{45-50\%}{\text{(Te)}} \end{cases}$$

1780   Reductive Amination with Sodium Cyanoborohydride: N,N-Dimethylcyclohexylamine

Richard F. Borch, School of Chemistry, University of Minneosta, Minneapolis, Minneosta 55455

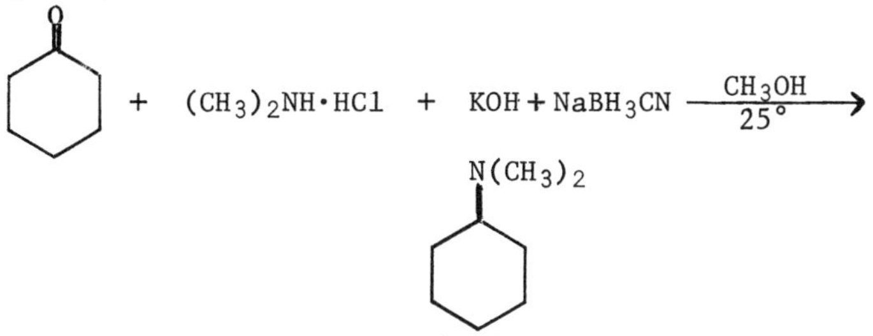

cyclohexanone + $(CH_3)_2NH \cdot HCl$ + $KOH + NaBH_3CN \xrightarrow[25°]{CH_3OH}$ N,N-dimethylcyclohexylamine

62-69%
(99.2% pure by vpc)

1781   Methyl Nitroacetate

Shonosuke Zen, Masao Koyama, and Shinkichi Koto, School of Pharmaceutical Sciences, Kitasato University, 130, Shirokane-Sankocho, Shiba, Minato-ku, Tokyo, Japan

$$CH_3NO_2 + KOH \xrightarrow[reflux]{H_2O} KO_2N=CHCOOK \xrightarrow[-15-25°]{CH_3OH\ H_2SO_4} O_2NCH_2COOCH_3$$

88%   83%
(73% overall)

1782 Dimethyl Nitrosuccinate and Dimethyl α-Nitroglutarate

Shonosuke Zen and Eisuke Kaji, College of Pharmaceutical Sciences, Kitasato University, Shirokane, Minato-ku, Tokyo, Japan

$$O_2NCH_2COOCH_3 + BrCH_2COOCH_3 \xrightarrow[\substack{CH_3CON(CH_3)_2 \\ 25°}]{\substack{CH_3ONa \\ CH_3OH}} \underset{\underset{CH_2COOCH_3}{|}}{O_2NCHCOOCH_3} \quad 66\%$$

$$Na^+ \; ^\ominus O_2N=CHCOOCH_3 + BrCH_2CH_2COOCH_3 \xrightarrow[\substack{25° \text{ then } 60°}]{CH_3CON(CH_3)_2} \underset{\underset{\underset{CH_2COOCH_3}{|}}{CH_2}}{\overset{O_2NCHCOOCH_3}{|}} \quad 88\%$$

1783 Diazoacetophenone

John N. Bridson and John Hooz, Department of Chemistry, The University of Alberta, Edmonton 7, Alberta, Canada

$$\phi COCl + CH_2N_2 + Et_3N \xrightarrow[-5° \text{ to } -10°]{Et_2O} \phi COCHN_2 + Et_3N \cdot HCl$$
$$\qquad\qquad\qquad\qquad\qquad\qquad\qquad 84\% \qquad 83\%$$

One-Step Synthesis of Benzyne from Aniline: 1,2,3,4-Tetraphenylnaphthalene

B. D. Baigrie, J. I. G. Cadogan, J. R. Mitchell, and J. T. Sharp, Department of Chemistry, West Mains Road, University of Edinburgh, Edinburgh EH 9 3JJ, Scotland, U. K.

1785   [18]Annulene (1,3,5,7,9,11,15,17-Cyclooctadecanonaene)

K. Stöckel and F. Sondheimer, Department of Chemistry, University College, London, 20 Gordon Street, London WC1H OAJ, England, U.K.

$$HC\equiv CCH_2CH_2C\equiv CH + (CH_3COO)_2Cu \cdot H_2O \xrightarrow[55°]{pyridine}$$

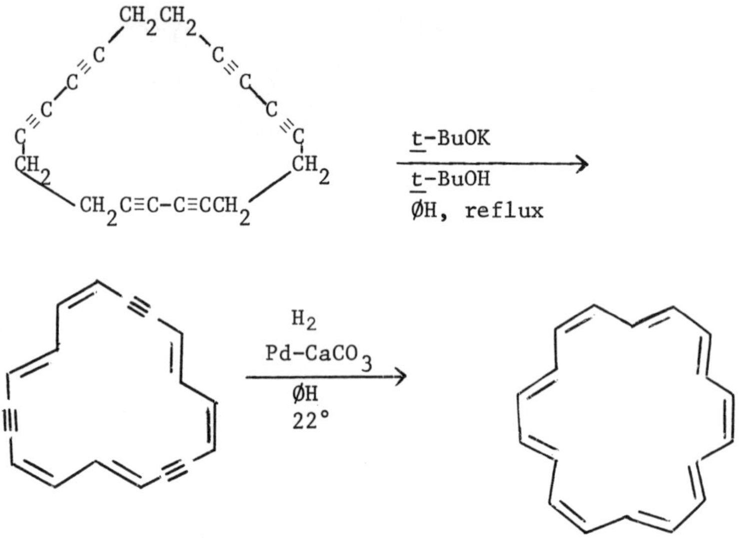

[cyclic intermediate with CH$_2$ groups and C≡C bonds] $\xrightarrow[\emptyset H, reflux]{t\text{-BuOK} \atop t\text{-BuOH}}$

[dehydro[18]annulene] $\xrightarrow[22°]{H_2 \atop Pd\text{-}CaCO_3 \atop \emptyset H}$ [[18]annulene]

2.6%   0.6%
(estimated)   overall

1786   Dideuteriodiazomethane

P. G. Gassman and W. J. Greenlee, Department of Chemistry, The Ohio State University, 140 West 18th Avenue, Columbus, Ohio 43210

$$CH_2N_2 + D_2O \xrightarrow[\substack{Et_2O \\ THF \\ 0°}]{\substack{NaOD \\ D_2O}} CD_2N_2$$

35-40% (based on N-nitrosomethylurea)
(98-99% D)

1787 3-Acetyl-2,4-dimethylfuran

P. D. Howes and C. J. M. Stirling, Department of Chemistry, University of North Wales, Bangor, Caernarvonshire, North Wales, U.K.

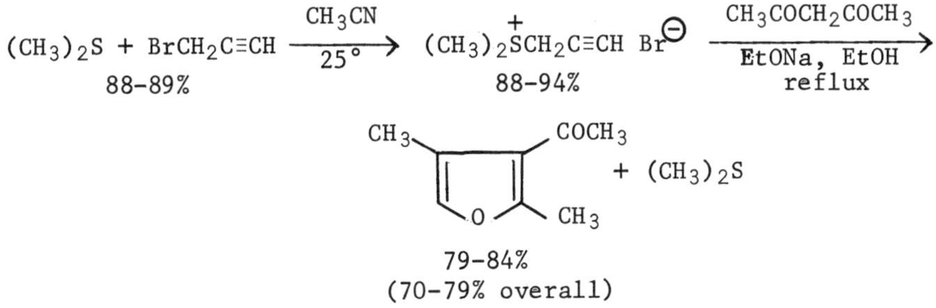

$$(CH_3)_2S + BrCH_2C\equiv CH \xrightarrow[25°]{CH_3CN} (CH_3)_2\overset{+}{S}CH_2C\equiv CH\ Br^{\ominus} \xrightarrow[\text{reflux}]{CH_3COCH_2COCH_3, \text{EtONa, EtOH}}$$

88–89% 88–94%

[3-acetyl-2,4-dimethylfuran] + $(CH_3)_2S$

79–84%
(70–79% overall)

1788 1,1-Bis(diethylaminomethyl)acetone

P. V. Ramani, J. P. John, and S. Swaminathan, Department of Organic Chemistry, University of Madras, Madras 25, India

$$2Et_2NH\cdot HCl + 2CH_2O + CH_3COCH_3 \xrightarrow[0-5°]{KOH, H_2O} (Et_2NCH_2)_2CHCOCH_3$$

50–56%

1789 1,3-Bis(methylthio)propene

Bruce W. Erickson, The Rockefeller University, New York, New York 10021

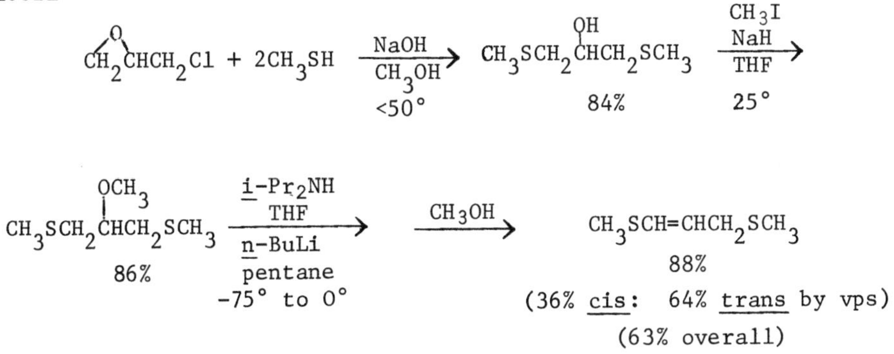

$$\underset{\triangle}{CH_2CHCH_2Cl} + 2CH_3SH \xrightarrow[<50°]{NaOH, CH_3OH} CH_3SCH_2\overset{OH}{CHCH_2}SCH_3 \xrightarrow[THF]{CH_3I, NaH} 25°$$

84%

$$CH_3SCH_2\overset{OCH_3}{CHCH_2}SCH_3 \xrightarrow[\text{n-BuLi, pentane, -75° to 0°}]{i-Pr_2NH, THF} \xrightarrow{CH_3OH} CH_3SCH=CHCH_2SCH_3$$

86%   88%
(36% cis: 64% trans by vpc)
(63% overall)

1790 3-(p-Chlorophenyl)-5-(p-methoxyphenyl)isoxazole

Matilda Perkins, Charles F. Beam, Morgan C. D. Dyer, and the late Charles R. Hauser, William Chandler Chemistry Laboratory, Lehigh University, Bethlehem, Pa. 18015

1791 4,4-Dimethyl-2-cyclohexen-1-one

Yihlin Chan and William W. Epstein, Department of Chemistry, University of Utah, Salt Lake City, Utah 84112

1792 Use of Silver Carbonate on Celite for Oxidative Coupling of Phenols

V. Balogh, M. Fétizon, and M. Golfier, Laboratoire de Synthese Organique, Ecole Polytechnique, 17, Rue Descartes-75, Paris 5ᵉ, France

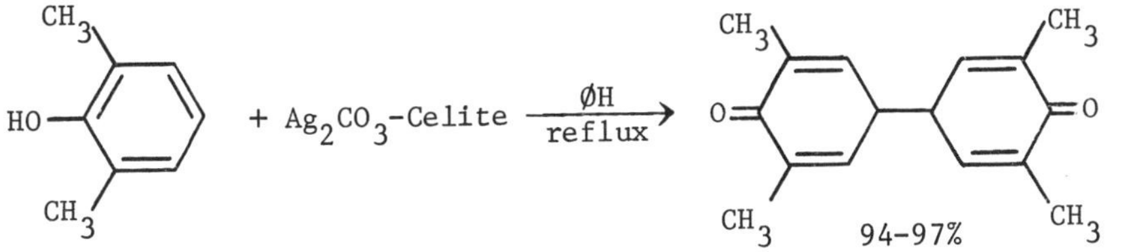

$$\text{2,6-dimethylphenol} + Ag_2CO_3\text{-Celite} \xrightarrow[\text{reflux}]{\phi H} \text{3,3',5,5'-tetramethyldiphenoquinone} \quad 94\text{-}97\%$$

1793 Acetylferrocene

George W. Gokel and Ivar K. Ugi, Department of Chemistry, University of Southern California, University Park, Los Angeles, Calif. 90007

$$\text{Ferrocene} + CH_3COCl \xrightarrow[0\text{-}5°]{\substack{AlCl_3 \\ CH_2Cl_2}} \xrightarrow[Na_2S_2O_4]{H_2O} \text{Acetylferrocene}$$

99% crude
85% pure

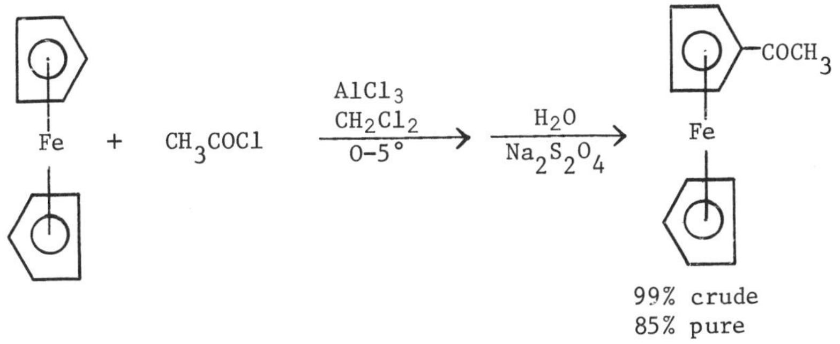

1794 1-Ferrocenylethanol

George W. Gokel and Ivar K. Ugi, Department of Chemistry, University of Southern California, University Park, Los Angeles, Calif. 90007

$$\text{Acetylferrocene} + Na^+AlH_2^-(OCH_2CH_2OCH_3)_2 \xrightarrow[50°]{\phi H} \xrightarrow[NH_4Cl]{H_2O} \text{1-Ferrocenylethanol}$$

85-90%

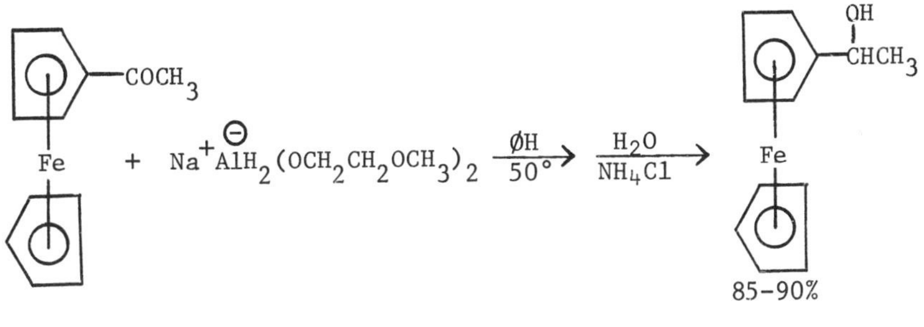

1795 R-(+)- and S-(-)-N,N-Dimethyl-1-ferrocenylethylamine

George W. Gokel and Ivar K. Ugi, Department of Chemistry, University of Southern California, University Park, Los Angeles, Calif. 90007

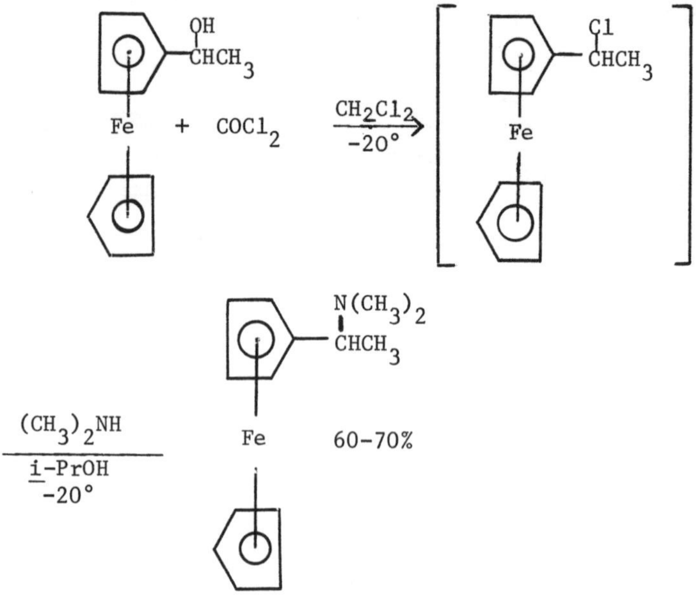

$$\text{Fc-CH(OH)CH}_3 + \text{COCl}_2 \xrightarrow[-20°]{\text{CH}_2\text{Cl}_2} [\text{Fc-CH(Cl)CH}_3]$$

$$\xrightarrow[\substack{\text{i-PrOH}\\-20°}]{(\text{CH}_3)_2\text{NH}} \text{Fc-CH(N(CH}_3)_2)\text{CH}_3 \quad 60\text{-}70\%$$

Subsequently resolved with R-(+)-tartaric acid.

1796 Dialkoxycarbonium Ion Salts as Alkylating Agents: N-Ethylbenzylamine

Richard F. Borch, Department of Chemistry, University of Minnesota, Minneapolis, Minn. 55455

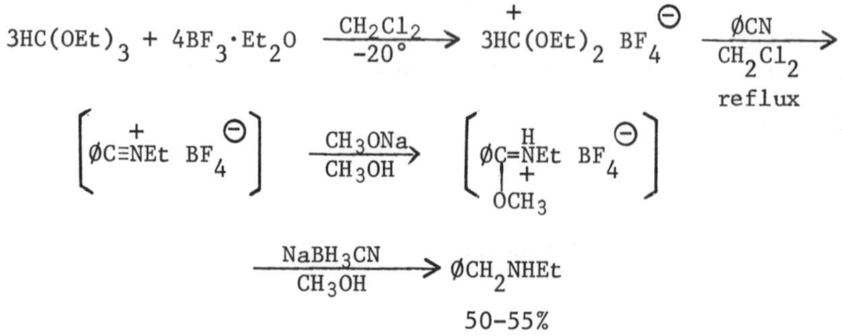

$$3\text{HC(OEt)}_3 + 4\text{BF}_3\cdot\text{Et}_2\text{O} \xrightarrow[-20°]{\text{CH}_2\text{Cl}_2} 3\text{HC(OEt)}_2{}^+ \text{BF}_4{}^- \xrightarrow[\text{CH}_2\text{Cl}_2]{\varnothing\text{CN}} [\varnothing\text{C}\equiv\text{NEt}\ \text{BF}_4{}^-]$$
reflux

$$\xrightarrow[\text{CH}_3\text{OH}]{\text{CH}_3\text{ONa}} \left[\varnothing\overset{\text{H}}{\underset{\overset{|}{\text{OCH}_3}}{\text{C}}}{=}\text{NEt}\ \text{BF}_4{}^-\right]$$

$$\xrightarrow[\text{CH}_3\text{OH}]{\text{NaBH}_3\text{CN}} \varnothing\text{CH}_2\text{NHEt}$$

50-55%

1797 **t-Butylcyanoketene**

Walter Weyler, Jr., Warren G. Duncan, Margo Beth Liewen, and Harold W. Moore, Department of Chemistry, University of California, Irvine, Irvine, Calif. 92664

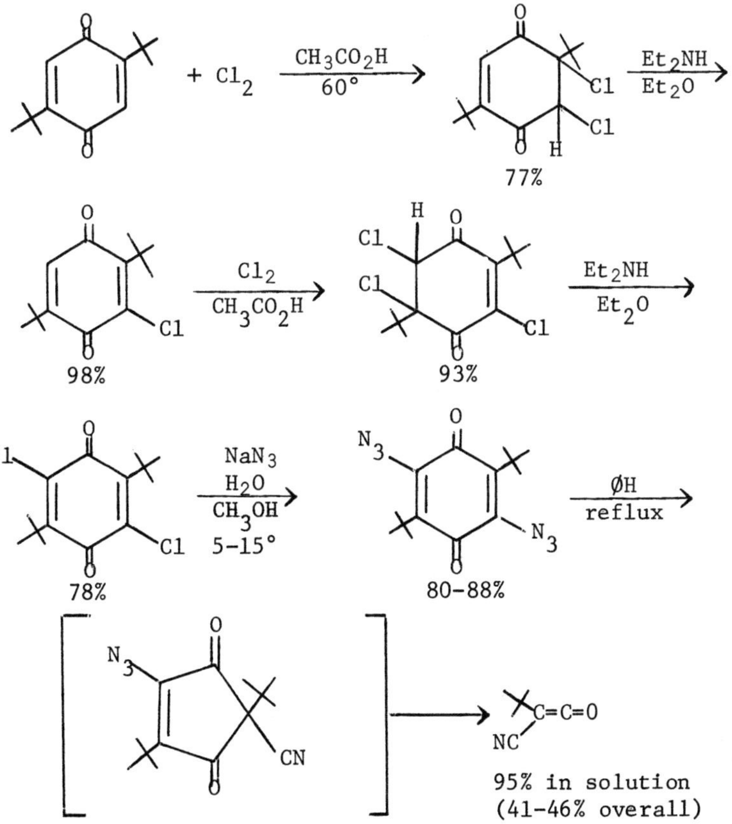

1798 **t-Butyl Azidoformate**

Koji Sakai and J.-P. Anselme, University of Massachusetts-Boston, 100 Arlington St., Boston, Mass. 02116

$$t\text{-BuOH} + COCl_2 \xrightarrow[25°]{Et_2O} t\text{-BuOCOCl} \xrightarrow[\substack{CHCl_3 \\ 0°}]{[(CH_3)_2N]_2C=NH_2^+ \; N_3^-} t\text{-BuOCON}_3 \quad 95\text{-}97\%$$

1799 2,2'-Bipyrimidine

T. R. Musgrave and P. A. Westcott, Department of Chemistry, Colorado State University, Fort Collins, Colo. 80521

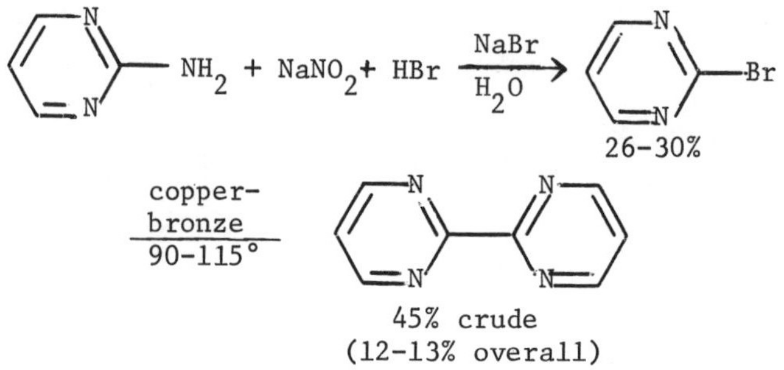

pyrimidine-NH$_2$ + NaNO$_2$ + HBr $\xrightarrow{\text{NaBr}, H_2O}$ pyrimidine-Br   26–30%

$\xrightarrow[90-115°]{\text{copper-bronze}}$ 2,2'-bipyrimidine

45% crude
(12–13% overall)

1800 p-Hydroxyphenylacetaldehyde and 4-Hydroxy-3-methoxyphenylacetaldehyde

Jay H. Robbins, Building 10, Room 12-N-238, National Institutes of Health, Bethesda, Maryland 20014

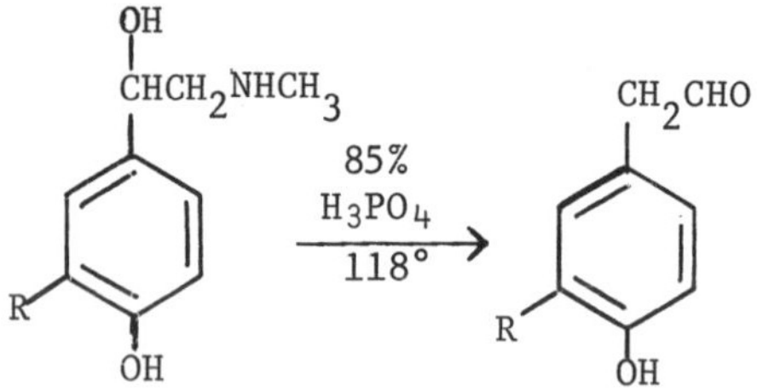

Ar-CH(OH)CH$_2$NHCH$_3$ $\xrightarrow[118°]{85\% \ H_3PO_4}$ Ar-CH$_2$CHO

Isolated as the NaHSO$_3$ addition products:
R=H   22% crude
R=OCH$_3$ 54% crude

1801 Vinyl Trifluoromethanesulfonates. I. From Acetylenes. cis and trans 2-Buten-2-yl Triflate

Peter J. Stang, Department of Chemistry, University of Utah, Salt Lake City, Utah 84112

CH$_3$C≡CCH$_3$ + CF$_3$SO$_3$H $\xrightarrow{-75° \text{ to } 0°}$ CH$_3$CH=C(CH$_3$)OSO$_2$CF$_3$ 55%

1802   Vinyl Trifluoromethanesulfonates. II. From Ketones.
       3-Methyl-2-buten-2-yl Triflate

Peter J. Stang and Thomas E. Dueber, Department of Chemistry,
University of Utah, Salt Lake City, Utah 84112

$$(CH_3)_2CHCOCH_3 + CF_3SO_3H \xrightarrow[-78 \text{ to } 25°]{\text{pyridine} \atop CCl_4} (CH_3)_2C=\underset{CH_3}{C}OSO_2CF_3 \quad 58\%$$

1803   Pinocarveol

J. K. Crandall and L. C. Crawley, Department of Chemistry, Indiana
University, Bloomington, Indiana 47401

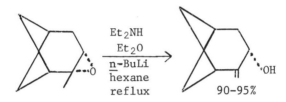

1804   Stereoselective Synthesis of Trisubstituted Double Bonds.
       Ethyl trans-4-Methyl-4,8-nonadienoate

Ronald I. Trust and Robert E. Ireland, Gates and Crellin
Laboratories of Chemistry, California Institute of Technology,
Pasadena, Calif. 91109

$$CH_2=CHCH_2CH_2Cl + Mg \xrightarrow[\text{reflux}]{Et_2O} \left[CH_2=CHCH_2CH_2MgCl\right] \xrightarrow[Et_2O \atop \text{reflux}]{CH_2=\overset{CH_3}{C}CHO}$$

$$CH_2=CHCH_2CH_2\underset{|}{\overset{OH}{C}}H-\underset{|}{\overset{CH_3}{C}}=CH_2 \xrightarrow[CH_3CH_2CO_2H \atop 138-140°]{CH_3C(OEt)_3} CH_2=CHCH_2CH_2\begin{array}{c}H\\ \diagdown\\ C=C\\ \diagup\end{array}\begin{array}{c}CH_2CH_2COOEt\\ \\ CH_3\end{array}$$

54-68%   83-88%
                                                 (45-60% overall)
                                                 (97% trans: 3% cis by nmr)

1805   N-Decylidenemethylamine

A. O. Bedenbaugh, W. A. Bergin, J. H. Bedenbaugh, and J. D. Adkins, Department of Chemistry, University of Southern Mississippi, Box 466, Southern Station, Hattiesburg, Miss. 39401

$$CH_3(CH_2)_8COOH + CH_3NH_2 + Li \xrightarrow[-78°]{} \xrightarrow[NH_2Cl]{H_2O}$$

$$\left[CH_3(CH_2)_8CH=NCH_3 \right] \xrightarrow[H_2O]{HCl} CH_3(CH_2)_8CHO$$
$$53-61\%$$

$$Et_2O \downarrow \underset{Pd-C}{H_2}$$

$$CH_3(CH_2)_8CH_2NHCH_3$$
$$68\%$$

1806   A General Synthesis of 4-Isoxazolecarboxylic Esters. Ethyl 3-Ethyl-5-methyl-4-isoxazolecarboxylate

John E. McMurry, Division of Natural Sciences, University of California–Santa Cruz, Santa Cruz, Calif. 95060

⟨NH⟩ + CH_3COCH_2COOEt $\xrightarrow[\text{reflux}]{\phi H}$ ⟨N-C(CH$_3$)=CHCOOEt⟩   98%

$\xrightarrow[\text{CHCl}_3, 0°]{CH_3CH_2CH_2NO_2,\ Et_3N,\ POCl_3}$ [isoxazole: CH_3CH_2-, COOEt, CH_3]   71%

(70% overall)

1807 Isoxazole Annelation Reaction. 1-Methyl-4,4a,5,6,7,8-hexahydronaphthalen-2(3H)-one

John E. McMurry, Division of Natural Sciences, University of California-Santa Cruz, Santa Cruz, Calif. 95060

1808 1-Phenyl-4-phosphorinanone

Theodore E. Snider, Don L. Morris, K. C. Srivastava, and K. D. Berlin, Department of Chemistry, Oklahoma State University, Stillwater, Okla. 74074

PhPH$_2$ + 2CH$_2$=CHCN $\xrightarrow[<35°]{\text{KOH} \atop \text{CH}_3\text{CN}}$ PhP(CH$_2$CH$_2$CN)$_2$ 75-85%

$\xrightarrow[\varnothing\text{CH}_3]{\text{t-BuOK}}$ reflux

[cyclic enamine with NH$_2$, CN, P-Ph] 90-95%

$\xrightarrow[\text{reflux}]{\text{HCl} \atop \text{H}_2\text{O}}$

[1-phenyl-4-phosphorinanone] 71-80% (48-65% overall)

1809 n-Butyl 1-Naphthyl Sulfide

J. S. Bradshaw and E. Y. Chen, Department of Chemistry 225 ESC, Brigham Young University, Provo, Utah 85601

1-ClC$_{10}$H$_7$ + HSCH$_2$CH$_2$CH$_2$CH$_3$ $\xrightarrow[\text{(CH}_3\text{)}_2\text{SO} \atop \text{reflux}]{\text{CH}_3\text{ONa}}$ 1-(SCH$_2$CH$_2$CH$_2$CH$_3$)C$_{10}$H$_7$ 70-80%

1810 Homogeneous Catalytic Hydrogenation: Dihydrocarvone

Robert E. Ireland and P. Bey, Gates and Crellin Laboratories of Chemistry, California Institute of Technology, Pasadena, Calif. 91109

$$\text{carvone} \xrightarrow[\substack{\varnothing H \\ 25°}]{\substack{H_2 \\ (\varnothing_3P)_3RhCl}} \text{dihydrocarvone} \quad 90\%$$

1811 2,3,4,5-Tetrahydropyridine

G. P. Claxton, L. Allen, and J. M. Grisar, Merrell-National Laboratories, Division of Richardson-Merrell Inc., Cincinnati, Ohio 45215

$$\text{piperidine} \xrightarrow[\substack{CH_3COOH \\ < 0°}]{\substack{Ca(OCl)_2 \\ H_2O}} \left[\text{N-Cl piperidine} \right] \xrightarrow[\substack{EtOH \\ reflux}]{KOH} \text{(trimer)}_3 \quad 42\text{-}64\%$$

1812 17β-Hydroxy-5-oxo-3,5-seco-4-norandrostane-3-carboxylic Acid

L. Milewich and L. R. Axelrod, Division of Biological Growth and Development, Southwest Foundation for Research and Education, P.O. Box 28147, San Antonio, Texas

$$\text{steroid-OCOCH}_3 \xrightarrow[\substack{H_2O \\ \underline{t}\text{-BuOH} \\ 35°}]{\substack{NaIO_4 \\ KMnO_4 \\ K_2CO_3}} \text{seco-acid} \quad 67\text{-}89\%$$

1813 Procedure for Reductive Cleavage of Allylic Alcohols or Acetates to Olefins

Irène Felkin, Institut de Chimie des Substances Naturelles, Centre National de la Recherche Scientifique, 91-Gif-Sur-Yvette, France

[structure with OH] $\xrightarrow{\substack{Zn-Hg \\ HCl \\ Et_2O \\ -15°}}$ [structure]

cis and trans   70%

1814 Acyloin Condensation in Which Trimethylchlorosilane is Used For Trapping Agent, With the Subsequent Hydrolysis Step. 1,2-Bistrimethylsilyloxycyclobutene and Succinoin

Jordan J. Bloomfield and Janice M. Nelke, Corporate Research Department, Monsanto Company, St. Louis, Mo. 63166

$$\begin{array}{c} CH_2COOEt \\ | \\ CH_2COOEt \end{array} + \underset{(or\ with\ K)}{Na} + ClSi(CH_3)_3 \xrightarrow[reflux]{\substack{\varnothing CH_3 \\ (or\ with \\ K, Et_2O)}}$$

[cyclobutene with two $OSi(CH_3)_3$ groups] $\xrightarrow[reflux]{CH_3OH}$ [cyclobutanone with OH]

65-80%   71-86%
(or with K, 78-93%)   (46-69% overall, or with K, 55-80%)

1815 Trimethylene Dithiotosylate

R. B. Woodward, I. J. Pachter, and Monte L. Scheinbaum*,
(*) Department of Chemistry, East Tennessee State University
Johnson City, Tenn. 37601

$$KOH + H_2S \xrightarrow[0°]{H_2O} KSH + H_2O$$

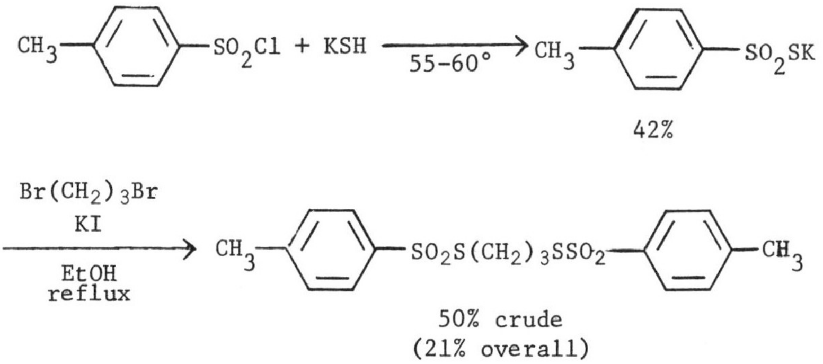

$$CH_3-C_6H_4-SO_2Cl + KSH \xrightarrow{55-60°} CH_3-C_6H_4-SO_2SK$$
42%

$$\xrightarrow[\text{EtOH reflux}]{Br(CH_2)_3Br, KI} CH_3-C_6H_4-SO_2S(CH_2)_3SSO_2-C_6H_4-CH_3$$

50% crude
(21% overall)

1816 2,2-Ethylenedithiocyclohexanone

R. B. Woodward, I. J. Pachter, and M. L. Scheinbaum*, (*) Department
of Chemistry, East Tennessee State University, Johnson City, Tenn.
37601

cyclohexanone + HCOOEt $\xrightarrow[\text{ØH} \\ 0-5°]{CH_3ONa}$ 2-(hydroxymethylene)cyclohexanone
70%

2-(hydroxymethylene)cyclohexanone + $CH_3-C_6H_4-SO_2S(CH_2)_2SSO_2-C_6H_4-CH_3$

$\xrightarrow[\text{reflux}]{CH_3COOK, CH_3OH}$ 2,2-ethylenedithiocyclohexanone

68%
(48% overall)

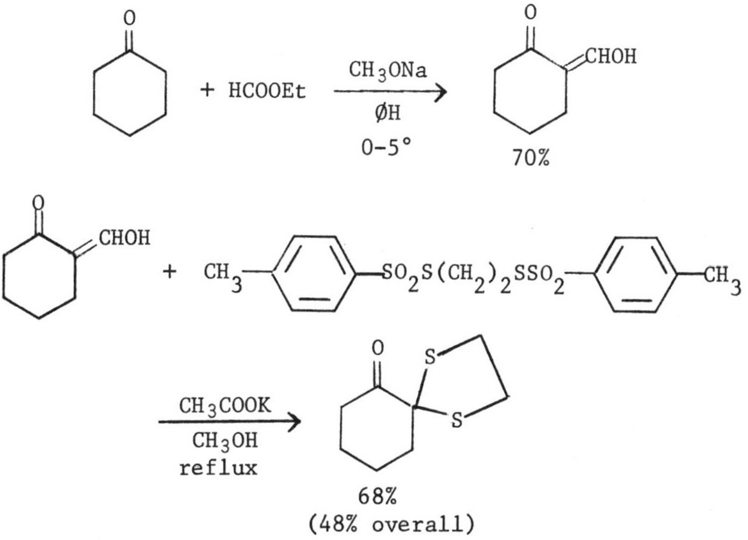

1817   2,2-Trimethylenedithiocyclohexanone

R. B. Woodward, I. J. Pachter, and M. L. Scheinbaum*, (*) Department
of Chemistry, East Tennessee State University, Johnson City, Tenn.
37601

cyclohexanone + pyrrolidine $\xrightarrow[\text{reflux}]{\phi H}$ 1-(cyclohex-1-enyl)pyrrolidine   "essentially quantitative"

1-(cyclohex-1-enyl)pyrrolidine + $CH_3\text{-}C_6H_4\text{-}SO_2S(CH_2)_3SSO_2\text{-}C_6H_4\text{-}CH_3$

$\xrightarrow[\text{reflux}]{\substack{Et_3N \\ CH_3CN}}$ $\xrightarrow[50°]{\substack{HCl \\ H_2O}}$ 2,2-trimethylenedithiocyclohexanone   37%

1818   Acylamidoalkyl Acetophenones from Substituted Phenethylamines:
2-(2-Acetamidoethyl)-4,5-dimethoxyacetophenone

A. Brossi, L. A. Dolan, and S. Teitel, Research Division, Hoffmann-
LaRoche, Inc., Nutley, N. J. 07110

3,4-dimethoxy-N-acetylphenethylamine $\xrightarrow[\text{reflux}]{\substack{POCl_3 \\ \phi CH_3}}$ 6,7-dimethoxy-1-methyl-3,4-dihydroisoquinoline   86-90%

$\xrightarrow[\substack{\text{pyridine} \\ 90\text{-}95°}]{(CH_3CO)_2O}$ 6,7-dimethoxy-2-acetyl-1-methylene-1,2,3,4-tetrahydroisoquinoline   72-77% $\xrightarrow[60\text{-}65°]{\substack{HCl \\ H_2O}}$

2-(2-acetamidoethyl)-4,5-dimethoxyacetophenone   91-93%
(56-64% overall)

1819   trans-1,2-Diacetylcyclopropane

Michael B. D'Amore and Robert G. Bergman, Gates and Crellin Laboratories of Chemistry, California Institute of Technology, Pasadena, Calif. 91109

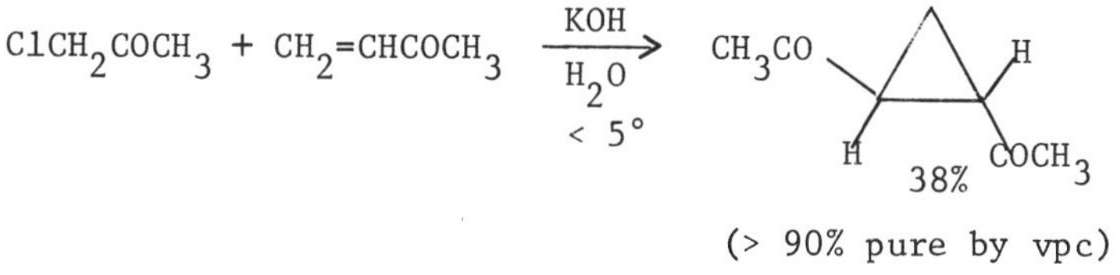

$$ClCH_2COCH_3 + CH_2=CHCOCH_3 \xrightarrow[\substack{H_2O \\ < 5°}]{KOH} CH_3CO\text{-cyclopropane-}COCH_3$$

38%

(> 90% pure by vpc)

1820   Orcinol Monomethyl Ether

R. N. Mirrington and G. I. Fentrill, Department of Organic Chemistry, University of Western Australia, Nedlands, Western Australia

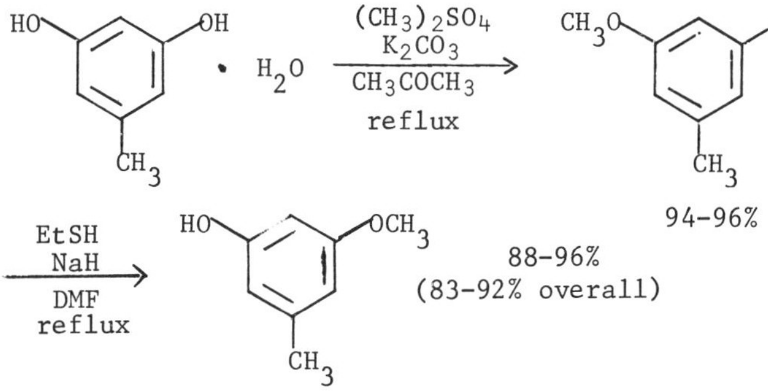

[Scheme: 3,5-dihydroxytoluene·H₂O → (CH₃)₂SO₄, K₂CO₃, CH₃COCH₃, reflux → 3,5-dimethoxytoluene (94-96%); then EtSH, NaH, DMF, reflux → 3-hydroxy-5-methoxytoluene, 88-96% (83-92% overall)]

1821   Syntheses Using Polymer-Supported Catalysts: Preparation of Dicyclopropylcarbinyl Ethyl Ether

D. C. Neckers and Robert L. Edgar, Department of Chemistry, The University of New Mexico, Albuquerque, New Mexico 87106

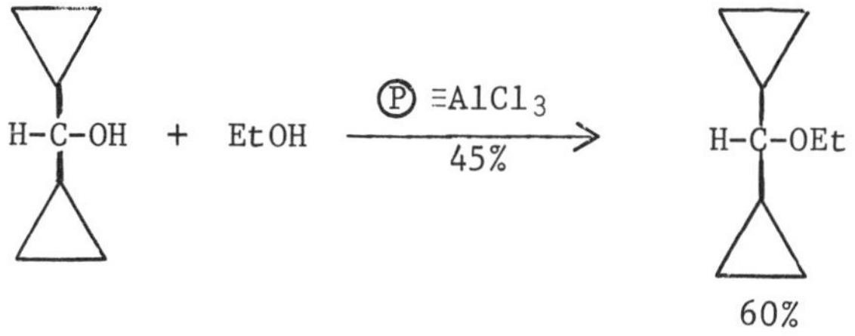

$$\text{(cyclopropyl)}_2\text{CH-OH} + \text{EtOH} \xrightarrow[45\%]{\text{Ⓟ}\equiv AlCl_3} \text{(cyclopropyl)}_2\text{CH-OEt}$$

60%

Ⓟ = styrene-divinylbenzene copolymer